BODY WEATHER

NOTES ON CHRONIC ILLNESS IN THE ANTHROPOCENE

LORRAINE BOISSONEAULT

BEACON PRESS
BOSTON

BEACON PRESS
24 Farnsworth Street
Boston, Massachusetts
www.beacon.org

Beacon Press books
are published under the auspices of
the Unitarian Universalist Association of Congregations.

© 2026 by Lorraine Boissoneault

All rights reserved
Printed in the United States of America

29 28 27 26 8 7 6 5 4 3 2 1

This book is printed on acid-free paper that meets the uncoated paper ANSI/NISO specifications for permanence as revised in 1992.

Text design and composition by Travis Cohen and Kim Arney

Excerpt from poem "Kindness." Printed here with permission of the author, Naomi Shihab Nye. From *Words Under the Words* (Portland, OR: Far Corner Books, 1995).

Certain portions of this book draw from earlier work that appeared in different contexts and have been modified and expanded for this publication:

Lorraine Boissoneault, "Finding a Face for My Invisible Illness," *Catapult*, July 26, 2021, https://magazine.catapult.co/people/stories/lorraine-boissoneault-essay-finding-face-for-my-invisible-illness-thyroid-hashimotos-goiter.

Lorraine Boissoneault, "Learning to Live with a Broken Heartbeat," *New Yorker*, July 30, 2022, https://www.newyorker.com/culture/personal-history/learning-to-live-with-a-broken-heartbeat.

Library of Congress Cataloging-in-Publication Data is available for this title.
ISBN: 978-0-8070-1755-5;
e-book: 978-0-8070-1756-2; audiobook: 978-0-8070-2326-6

The authorized representative in the EU for product safety and compliance is Easy Access System Europe 16879218, Mustamäe tee 50, 10621 Tallinn, Estonia: http://beacon.org/eu-contact.

For Kevin,
who's been there
for all of it and
never looked away.

CONTENTS

PART FOUR

LANDSLIDES—GUTS—GRIEF

PART FIVE

FIRE—JOINTS—RADICAL LOVE

PROLOGUE

WEATHER WALKS

It started with clouds. Not in the way of some Gothic novel signaling a character's mindset or foreshadowing tragedy through claustrophobic grayness, nor like the whimsical game of looking for shapes in the white fluff of water vapor. No, I was looking up at the sky to, paradoxically, ground myself in my body. Call it the environmental form of mirroring: I wanted to assess how the atmosphere was feeling at the same time I noted the feelings within my "bodymind," to see the ways we overlapped and diverged and interacted. I was building a world map where the boundaries between myself and everything else were purposefully thin, sometimes permeable. I needed the reminder that who I am extends beyond the barrier of my skin, beyond even the invisible clouds of breath and microbes and heat my body sheds at every moment. I am a tiny piece of a planetary system, influenced and influencing.

And so I walked and tried to study the clouds, giving them their scientific names (cumulus, nimbostratus, cirrus) while I assessed the weather within myself. Where the aches, the stabbing pains, the heat, the nausea? Where the comfort, the ease, the strength? Some days I walked in rain; some days I slipped on ice; some days I used a cane; some days my body felt soft as the summer air. Whenever I returned home, however long or short the trek, I jotted my observations: the sky and the temperature and my body, a few small words to signify these vast mysterious forces.

In a way, that's what it feels like I'm doing here, albeit with more than a few words. I'm not a meteorologist reporting on the day's weather and the ten-day forecast, nor am I a scientist probing Earth's core and its atmospheric currents to understand the larger picture of climate. I'm a writer in a sick body on a sickening planet, and I'm making both of us into symbols, characters in a vast narrative, in hopes that together we equal more than the sum of our parts—that in pairing up, we'll have more to say than "you're all doomed." And maybe this endeavor risks overemphasizing my individual experience, but you'll have to bear with me as I fumble my way forward. I'm making this up as I go along.

I started thinking about the similarities between meteorology and medicine when my partner asked about my daily "body weather." For most of my life I've lived in a humid continental climate with four distinct seasons. The weather changes daily, sometimes hourly. Predictions must be taken with a grain of salt. As I got sicker over the course of my twenties and thirties, my body created its own unpredictability. It became hard to know what each new day would bring. I fought it at first, sometimes literally; during boxing classes, I pictured my body as the punching bag, hoping the energy of my kicks and jabs would subdue everything going wrong inside me at a cellular level. Why was my body so determined to hurt me? I was angry at what was happening, angry at the way my health prevented me from making plans, imagining a future. And I was sad and scared and sometimes in denial about all of it. I was grieving. I was feeling the same emotions for my body that I felt for the planet as it contorted under the ravages of climate change.

Others with chronic illnesses and disabilities have expressed their own awareness of the overlap between their bodyminds and the weather.[1] The pressure change of storms can exacerbate migraines for some, worsen the vertigo of Ménière's disease for others. Dry, cold weather can trigger a flare-up of psoriasis. The lack of daylight in winter is associated with seasonal affective disorder, a form of depression. As poet Polly Atkin writes, "The body has its own weather, its own thunderstorms and floods, its heatwaves and hurricanes, its own natural disasters. It's enough to be at the mercy of the inner meteorology, never mind the outer. But this is how it is. The skin is not a dome habitat, sealed against climate. The weather outside alters the weather within."[2]

There's always some danger in reducing the experience of illness to a symbol, of using language in a way that obscures rather than illuminates. As Susan Sontag noted, "Nothing is more punitive than to give a disease a meaning—that meaning being invariably a moralistic one. Any important disease whose physical etiology is not understood, and for which treatment is ineffectual, tends to be awash in significance."[3] This is frequently the case with autoimmune diseases or illnesses like POTS (postural orthostatic tachycardia syndrome) and fibromyalgia and myalgic encephalomyelitis. Because their origins are difficult to ascertain, blame is laid all too often on the patient.

My goal isn't necessarily to assign meaning. I'm more interested in drawing connections and sharing experiences. But the experience of illness and climate change can both be strangely hard to convey. Maybe it's the enormity of them, the sense of being stranded in an entirely new world without so much as a guidebook. It reminds me of the work some researchers are undertaking into the "exposome," all the non-genetic exposures and interactions humans have with the environment over their lifetime: everything from the food we eat to the viruses we catch, our social interactions, and the chemicals we come in contact with. The concept was created as a play on "genome," and it yields a dizzying array of data, all to parse the way complex disease develops.[4] It's such a vast undertaking that we don't yet have the computing power to see the big picture.

The factors contributing to climate change—fossil fuel use, habitat destruction, pollution—are also making people sick. What we do to the planet, we do to ourselves. But these are big, ominous, almost ineffable facts. They're hard swallow, let alone digest. Only in bringing myself back into the body have I been able to grapple with the enormity of climate change and permanent illness. It isn't enough to think about them, to find statistics and watch footage of disasters and read studies in various academic journals. Living with personal and planetary crisis has required *feeling*, both physiologically and emotionally. How I feel the weather and how I feel about the weather; how I feel my body and how I feel about my body. And it was these feelings that formed a pathway through the wildness of unfolding catastrophe. They helped me go deep and wide: focusing inward, then turning my focus out onto the world.

In her memoir on surviving a debilitating autoimmune disease, Sarah Manguso writes, "Those who claim to write about something larger and more significant than the self sometimes fail to comprehend the dimensions of a self."[5] I'm not sure I understood all the dimensions of myself before I took the time to explore, deeply, my relationship to my body and my body's relationship to the world. Chronic illness and disability have been the hardest experiences in my life. They cracked me open, pumped me full of steroids and anger and hormones and grief and immunosuppressants and suffering. Manguso again: "This is suffering's lesson: *pay attention*. The important part might come in a form you do not recognize."[6]

When I couldn't bear to pay attention to my body anymore, I paid attention to the clouds. The air. The cycles of freeze and thaw and rain and drought, buds on the trees and cicadas tunneling out of the earth, the smell of storms and fire and mud and rain. And once I'd attended to the world's weather, my body's weather was easier to live with.

The stages of grief never really end; they're a cycle. Weather doesn't go away either. It shows different faces, realigning water and carbon and nitrogen. It responds to human activity in increasingly ferocious fashion. The weather demands to be felt. How much we're willing to feel it—the physicality of climate change as much as the emotions it engenders—will determine how well we survive what's coming next. No body is an island.[7]

PART ONE

TEMPERATURE—THYROID—DENIAL

WHEN THE TEMPERATURE HAS TEETH

Temperature is the backbone of weather, a core component of cellular function, and thus a keystone of life. It's so obvious and omnipresent a phenomenon that even if we check the forecast daily, it's possible to forget how deeply our lives are shaped by the amount of solar radiation we receive wherever we reside. The presence or absence of heat is a kind of alchemy, creating rain, sleet, fog, snow, hail, clouds, storms. It dictates whether bodies of water are liquid or solid, outside and within us. Ice isn't limited to winter ponds and Earth's two poles; frostbite is the freezing of flesh, tissues hardening and cell membranes potentially bursting. Temperature ties all animals, plants, air, and rocks into an unpredictable orchestra, the notes and rhythms changing even as we do our best to place predictions on them.

The year I was born, my birthplace of Toledo, Ohio, experienced a high of 96 degrees Fahrenheit and a low of minus 19 degrees. Such annual swings delineate the seasons around the Great Lakes. They marked as consistent a rhythm throughout my childhood as the academic calendar, even when I moved forty miles east, away from the city to a much smaller town. From the earliest part of my life, I understood there were more than five senses. How we feel temperature through nerve cells that reach to the outer layer of our skin intimately shapes the way we experience the world. Being too hot or too cold pushes us to seek relief. It influences our fashion, our architecture, what we eat, when and where we sleep. Though we rarely think about it in such terms, the need to maintain a stable core temperature is as crucial as our need for food and water.

I grew up near the shores of Lake Erie. Winters warped the streets and crusted the soil. Snowfall started in November and often lasted till April. Even with the bitterness of icy nights, winter was a season I loved. Getting bundled into snow pants and mittens and my hat to go sledding, the wind scraping my cheeks and snow sliding up the back of my jacket if I rolled off the sled at high speed. My wet hair clumping into icicles after swim practice at the YMCA. Hunkering down in the narrow cabin of an iceboat as the sails caught the wind and the blades vibrated as they cut tracks in the frozen lake. The way the outdoors smelled of pine and freezer burn after it snowed, and how my skin prickled for a few minutes when I first returned inside.

The summer came with its own patterns and sensations. Long, hot days started with milk and cereal and ended with fireflies blinking across grassy lawns, sometimes trapped in jars to illuminate my warm bedroom, devoid of air conditioning. I rode my bike to feel a breeze, dashed through sprinklers to rinse off sweat, accumulated itchy grass clippings around my ankles and toes. Summer meant sailing on Lake Erie and visiting my cousins who lived near the beach and riding roller coasters at Cedar Point amusement park.

Because my family favored outdoor sports, I grew up in close contact with the elements. The intensity of heat and cold wasn't something to avoid but a series of sensory experiences that could be modified. For the first part of my life, it was easy to ignore the profound fragility of humans living amid a fluctuating temperature, because I had a home and seasonally appropriate clothes. A changing temperature was no threat.

When does ignorance cross over into denial?

In Girl Scouts, I learned that people died of hypothermia or heatstroke. Nature was not uniformly benevolent; humans are animals like all others, with perhaps more vulnerable soft parts than most. But I trusted the solidity of my body and my understanding of the environment. They seemed in accord with each other. I was brought aboard my family's sailboat before I could walk; I learned to swim so early in life that I have no memory of ever *not* knowing the skill. I was swimming competitively by age eight and sailing competitively by age eleven.

At the time, swimming was my preference. Pools are controlled environments. The water temperature varied from one YMCA to another, but there was never a risk of freezing. Even if the air teased goosebumps across my wet skin after finishing an event, I could go stand under the hot spray of showers with other swimmers. There were towels and robes and blankets to wrap ourselves in. There was always food to snack on: hot dogs and walking tacos and pizza fueling our metabolisms, providing energy and heat.

Sailing was another story. Regatta season began in late May, when Lake Erie's water has an average temperature of 52°F. The first craft I ever raced on my own was a seven-foot-long boat called an Opti. The thing was shaped like a tub, eager to collect water. Two red buoyancy bags were strapped down at the front of the boat, and God help you if they leaked air. With bad bags, it was entirely possible to sink the boat, no matter how much water you bailed out (as I learned firsthand during one practice session). The single sail sat low enough above the hull that I was constantly getting banged in the head by it. I hated Optis because I was always tall for my age. The boat felt too small and cramped. But that's the vessel you had to race first if you wanted to graduate to bigger boats as a teenager.

One of the first regattas of the summer was at Jolly Roger Sailing Club in Toledo. I was dressed in warm gear and my life jacket, nervous but excited. The wind was brisk yet manageable—until I capsized and got soaked. It took some time to bail out the boat after I'd righted it. One of the race committee powerboats came over to ask if I wanted to go in. I was cold, but not freezing, and I didn't want to give up. I told them I'd keep racing. Did I capsize a second time? I might have. The memory of that day has grown hazy with time. What I remember is being wet and cold and thoroughly tired of my boat before the race finished. But I managed to win fourth place, earning my first sailing award. And as at the swim meets, there was plenty of tasty food when we went back to shore, and towels and hot showers—so many ways to get warm again. I wasn't sure I could say I had *fun*, but I'd proven my grit and determination. I'd proven what my body could do.

I was never the fastest or best at either sailing or swimming, could never predict how well I might do ahead of any race. But even as I grew taller, wrestled with puberty, made new friends at different schools, I maintained a fundamental confidence in my body's baseline abilities. If I got too hot, I'd cool back down. Too cold, and I'd warm up. Too tired, and I'd sleep to awake refreshed. It wasn't so much that I thought myself indestructible as it was a belief in my body's predictability. When I decided to go on an eleven-month exchange program to France at age sixteen, I assumed that no matter how much the cultural and linguistic trappings of my life daily changed, my body would behave as it always had.

And yes, the sailboats and wind and waves *were* familiar in France. I had to adapt to tides tugging Le Trieux in and out, blending sweet river water with salt from the English Channel ("the Sleeve," as I learned to call it in French). Here off the coast of Paimpol, I learned the smell of brackish water, saline splashes burning my eyes and throat. Low tide revealed rotting mud at the shore's edge, boats listing on their sides like beached whales. I thought, initially, that even with all these environmental changes, the biggest difficulty remained my limited understanding of French. Seated next to my host sister aboard a slow-turning Dart catamaran, I barely understood when she asked me to pull the sail in. I'd lived in France for one month.

The day we capsized was early fall, the sun breaking into golden panes on the water. A brisk wind made ruffles on the shoreline and sent our boat skittering. I thrilled at the speed even as I suspected it might end in disaster. We lifted onto one hull, straining to push our weight over the high edge of the tilting catamaran to counterbalance the force of the wind—and then we were over.

Plunging into 48-degree water sent daggers through my liver, my spleen, my stomach and lungs. I'd worn sweatpants and a sweatshirt and rain jacket, terrible sailing gear, but all I had in this country. Now the cotton turned heavy and sodden, doing little to trap my body heat against my skin. Gasping, hands going numb, I did my best to assist in righting our vessel, a process I'd achieved dozens of times on different boats in warmer conditions. Now it took too many minutes; once back atop the mesh trampoline spread between the boat's two hulls, I shook. Water poured from the sail, dripping over my soaked scalp. Wind wicked away every bit of warmth my heart managed to pump through my veins.

But the session wasn't over and my host sister didn't want to return to shore early. In her wetsuit and spray top, the water was no more than a nuisance. "J'ai froid," I tried to explain when the lines slid through my fingers. "J'ai froid," I reiterated as the cold assaulted me, penetrating so deeply it began feeling hard to breathe. "J'ai froid," I said, because I didn't know the words to explain that this was worse than being chilly, that I was developing hypothermia and desperately needed to get warm again. All I could say, again and again, was that I was cold. Such a mild word for the leadenness of my body, the nausea gnawing my stomach, the fog penetrating my thoughts.

When we finally reached shore, I stumbled to the showers, hoping warmth would penetrate my frozen core. In the mirror my lips showed blue, but everywhere else the hot water had turned my skin a boiled red. I waited fifteen minutes to feel less cold but only started shaking again. I wrapped myself in dry clothes for the drive home, then headed to my warm bed, even though it was only midday. The very next week my host family took me shopping for a wetsuit.

Nothing else came of that brief brush with hypothermia. Not at first. But when I try to track my body's new behavior in relation to temperature, it's this moment that grabs me. What had changed in the five years between my cold capsizing in Lake Erie and my slightly colder dunk in the River Trieux? Was it only a matter of a few degrees difference in water temperature, the heat my body managed to create in one outfit or another?

That year in France, I felt myself become a new person. But the source of those changes may have been more than adaptations to life in another country. From lymph and blood came the agents of metamorphosis, a process of which I could have no idea. Long before any tests were run, any diagnosis made, I think I was on my way to illness. I think that's when my body began to devour itself.

If any form of denial was at work in me, it's the one most of us spend our days employing: denying that significant changes to our body, our life, our world could happen at any moment. This omnipresent denial is, as a friend put it to me, "what gets us through the day [without] thinking about the abyss." Yes, this was how I lived. The patterns of my life changed, but my body's rhythms would remain predictable for years to come. I denied the possibility of disease because I had no idea that something physiological had changed in me. Not yet.

Humanity's relationship to temperature could be read as a charming example of perseverance or as tragic hubris, depending on how you view our species' impact on the planet. Some two million years ago, our hominin forebears lost most of their insulating fur.[1] At some point, they also gained an increased ability to sweat.[2] Thanks to these biological changes, determined bipeds were able to travel great distances by foot, beyond Africa and into Eurasia. Other animals who traverse vast latitudes, from migratory birds to humpback whales, rely on their physiology to buffer them from geographic thermal shifts. The genus *Homo*? We experimented with fire, clothing, and shelter. We exploited whatever environments we settled, fitting ourselves into the most unpredictable of niches. From arid savanna to high-altitude plateaus, our ingenuity and ability to mold our bodies to many environments has made us remarkably adaptable.

This is an especially impressive achievement when you understand how our bodies produce and respond to heat. Humans tend to think of temperature as a line on a thermometer, mercury rising as the sun beats down. But this is an oversimplification of the real way we feel and produce heat and cold. Humans and other mammals are endotherms; our bodies maintain relatively constant temperatures regardless of the external temperature. If it's cold, specific areas of the brain trigger shivering.[3] In heat, the hypothalamus (a structure in the brain that manages blood pressure, body temperature, hunger and thirst, mood, sleep, and sex drive) triggers blood vessels to dilate, causing the sweat response. We live in a relatively small window of tolerance; if our body temperature moves a few degrees above or below 98.6, whether due to fevers from infection or hypothermia from exposure, we begin to feel uncomfortable. A few more degrees warmer and we die.[4] Hypothermia, while dangerous, seems to offer a slightly wider range of survivability: the coldest core temperature a person has ever survived was 16°C/60.8°F.[5] We are endothermic homeotherms, blessed with the ability to generate our own heat, cursed to live in a narrow band of safety and comfort.

Ectothermic reptiles and amphibians, on the other hand, are often called "cold-blooded," for their inability to produce high levels of body heat. They outsource that work to the environment.[6] Because of this, they are also poikilotherms, meaning they have a variable body temperature. Some species can handle freezing temperatures, while others manage to get as hot as 47.5°C/117.5°F without dying (though they are still vulnerable to the increasing temperatures of climate change).[7]

But certain endothermic species show similarities to those in the poikilotherm clan; their body temperatures fluctuate widely without any lasting damage to the animal. Black bears' body temperatures drop during winter hibernation. Two-toed sloths are "facultative poikilotherms," their body temperature varying by up to 10°C, which allows them to obtain all the energy they need from a basic diet of leaves rather than foraging higher-calorie foods like fruit. And naked mole rats are heavily reliant on external temperatures to maintain their own body temperature, making them mammalian poikilotherms.[8]

All vertebrate life forms on Earth have their own mechanisms for staying warm and cooling down, but the body systems of most ectothermic poikilotherms are so unlike those of my own that I have a hard time imagining what the world must feel like in their pebbled or mucosal skin. Our lineages diverged three hundred million years ago; our ways of sensing the world have had so much time to differentiate.[9] Wood frogs freeze solid in the winter, their organs encased in ice but protected from cellular damage by a combination of glucose and urea; pit vipers sense infrared radiation from facial pit organs; some lizards detect scent with their tongues.

A bear though—a bear's experiences feel within reach of human comprehension, despite their ability to hibernate. These sometimes-bipedal, mostly omnivorous creatures can be curious or aggressive or scared. We've made them mascots, given them speech in cartoons, spent years studying their behavior, sometimes to our own detriment. When it's time for brown bears to hibernate, their metabolisms change, their thyroid hormones dropping to such low levels that their metabolism is down to 25 percent of its active state.[10]

What does it feel like, I want to ask, when you've grown large in the autumn and must lumber toward the overwinter den? Does the cold lose its bite when your body has pads of fat to ignite? Or is hibernation an annual exercise in painful self-destruction? And how do they remember, when the dormant season is over, how to start back up again, eating and mating and hiding from humans who sprawl, ant-like, across the same mountains where the bears have just spent a solitary winter?

I once saw a skinny bear on a steep hill while driving through Yellowstone National Park in the summer. The bear chewed some vegetation, maybe cow parsnip or thimbleberry. It didn't look capable of surviving a long winter, let alone the boom time of summer. Park rangers warned of the dangers to both humans and bears if our species came too close in proximity with each other. But there's no way to warn the bears of coming changes to the climate caused by human behavior—new precipitation patterns and vaster wildfires and warmer winters. It's possible the bears already know; they've started coming out of hibernation a full month early in zoos, and scientists have noted that wild populations are also changing their hibernation behavior.[11] I think maybe they're better at sensing these changes in their bodies than we are. Maybe it's because they can't be in denial; they feel the change, and they respond.

From my human perspective, it happened slowly: five or six years of subtly increasing sensitivity that started during my teenage years. A short walk outside in below-zero windchill would leave me shivering for hours. An underheated bedroom meant not falling asleep. Going out to bars in January would often require a hot shower or heating pad upon return, even when I eschewed a dress to cocoon myself in jeans and a parka. In classrooms and cafés, my hands felt like ice. Winter was no longer my favorite season.

Summer offered no reprieve. The heat astonished me, as if I'd spent years not noticing how brutal sunshine can be. I'd grow dizzy and weak but barely perspire. This bodily quirk was most notable in extreme situations. Running for miles on an unshaded path, then vomiting all the water I drank. Taking ten minutes longer to start sweating in a sauna when everyone around me was drenched by their perspiration. The temperature now had teeth, and no matter how I dressed it was impossible to avoid being bitten.

On a geological scale, the shift in my body was sudden. I learned this new perspective on time thanks to a sudden fascination with earth sciences. My first year of college I signed up for an Intro to Geology class because it fit into my schedule and filled a science requirement. Seated among hundreds of students in a large auditorium, I discovered that our planet could be understood as a series of stories being constantly rewritten, leaving a palimpsest in each layer. Fossils have fascinated us for much of human history, all the way back to the Romans making monsters of extinct animals, but rocks can chronicle epochal time even when they hold no obvious remains.[12]

Our planet's distance from the sun is a part of how it came to be habitable, but this is only one factor in the delicate balance of temperature control. So much depends on the interchange of energy between land and sky.[13] Early in its 4.56-billion-year history, Earth may have been as hot as 400°F, a temperature far too high for humans to survive.[14] During the Hadean Eon, Earth's magma ocean solidified; water covered the surface and landmasses formed.[15] Greenhouse gases were trapped within rock, buried under the sea. Over time Earth's surface cracked into a great array of shapes—progressing from Gondwana to supercontinent Pangaea to the configuration of continents that we live upon today.

As land and ocean engaged in their operatic struggle, the atmosphere etched its impact in subtler ways. It left traces in shellfish, in fossilized pollen.[16] Scientists now read these buried signs like a medium channeling the past. From their work, we know that Earth has cycled through many temperature variations.[17] There have been eons of ice and periods of high water. The interplay of hot and cold, sun and snow turned solid rocks into a tapestry.

It's easy to admire the endurance of rocks if you know nothing of geology; now I began to see the extent to which the atmosphere could break them down. Rocks are more sensitive than we imagine them to be. They respond to the world in ways we don't notice without microscopes or special knowledge: crumbling under ice, cracking in the heat, dissolving when atmospheric carbon dioxide concentrations make rain more acidic. The Great Lakes I'd grown up on were proof of stone's malleability; only twenty thousand years ago, retreating glaciers gouged deep rifts in the shale, limestone, and sandstone, then filled the cavernous holes with meltwater to form the lakes.[18]

I learned all this as my atmospheric sensitivity changed. My response to temperature was the most obvious warning sign, since I came in daily contact with the outdoors. But there were other symptoms. I went through episodes of exhaustion so intense that I visited the doctor anticipating a diagnosis of mononucleosis; I never tested positive. But I fell ill with plenty of other infectious diseases. A cold morphed into bronchitis, complete with fever and a monthlong cough, finished off by laryngitis. I caught strep throat during a summer program in China, had an allergic reaction to my antibiotics three days into treatment, then got strep a second time. I had so many urinary tract infections that I kept a stash of Azo pain relievers. Friends joked that I was a "cesspool," liable to come down with any bug roving the hallways of our dorm.

I resented being thought of as the sick one. I'd fancied myself the epitome of health since childhood. I'd had sports injuries and stitches, chickenpox and the flu, but nothing that ever seemed out of the ordinary. Elementary school classmates with asthma plus one fictional character with type 1 diabetes (Stacey McGill of *The Baby-Sitters Club*) were my only contact with childhood chronic illness. Though I no longer saw myself as a child in my late teens and early twenties, society had me convinced that I was too young to have health problems. Because there were no obvious answers, I ignored the possibility that there might be some overarching issue. My skin, my weight, my nails all looked the same. I denied that anything had changed because there was nothing to see, and because something as important as health was supposed to be in my control.[19] I should've realized from everything I'd learned about the planet that there's only so much the bare eye can discern. Invisible forces play a fundamental role in maintaining biological and atmospheric equilibrium.

French mathematician Joseph Fourier spent years investigating the mathematics of heat, and in 1824 published an article considering the role of the atmosphere in keeping Earth warm.[20] By 1896, Swedish physical chemist Svante Arrhenius modeled the way that increases or decreases in atmospheric carbon dioxide could change the temperature of the planet. The calculations were time-consuming, according to Arrhenius: "unbelievable that so trifling a matter has cost me a full year."[21] At the end of his work, he was able to assert that doubling the amount of CO_2 in the atmosphere would raise Earth's temperature by 3 to 4 degrees Celsius. Arrhenius was aware that human coal-burning added carbon dioxide to the atmosphere, but this didn't worry him. He estimated it would take at least five hundred years before atmospheric CO_2 doubled, at which point humans would probably be enjoying the warmer weather.

Even with British colonists following the Biblical mandate that man "fill the earth and subdue it," remaking the heavens must have seemed an impossibility. How much can one species do to such a vast planet? Yet only a century after Arrhenius's calculations, CO_2 levels had already increased by more than 20 percent, owing to an ever-growing reliance on fossil fuels.[22] We needed nothing like five centuries to transform this layer of Earth's skin and sky.

Climate science was yet another element of my geology education. People like to denigrate weather as the basest of subjects for small talk, but I always thought this was doing our atmosphere a disservice. Talking about the weather meant acknowledging that we all breathe the same air, while also seeing the multitude of ways in which this air can change from one place to the next, one moment to another. And couldn't people see that the climate was changing, from the weird weather patterns to the faster breakdown of rocks?[23] Though I hadn't learned much about it in high school, the science wasn't new. The *New York Times* ran a front-page story on climate change in 1983, and the Intergovernmental Panel on Climate Change was established in 1988, a year before I was born.[24]

Despite the mountain of scientific data supporting the facts of climate change, I knew plenty who denied its reality.[25] As heatwaves crisped countries, as the polar vortex dipped disturbingly far south, many dismissed the strangenesses as outliers. I didn't understand the ability to act is if nothing were wrong. The more I learned, the more my future on the planet felt precarious.

But I could still trust my body; of that I remained certain. I might be more sensitive to the temperature, but it was an insignificant change. Temperature sensitivity was no more worrying than the fact that I'd already sprouted a few gray hairs at the ripe age of twenty. Maybe I got sick more often than others and had an array of strange health crises throughout my four years of college in Ohio—a ruptured ovarian cyst, a recurrent heart arrhythmia. They didn't mean I was *sick*. They weren't part of a pattern. I told myself for years, *Nothing is wrong*. Until there came a point at which absolute denial was no longer possible.

I moved to New York City the summer of 2012, happy to be done with my undergrad education in Ohio, eager to start grad school and move in with my partner. Both of us spent most of our childhood in a town of six thousand people. Neither of us had ever lived in a city larger than three hundred thousand people for more than a few months, nor did we know anyone in the area. But he was starting a job and I'd be at Columbia University; we'd find our footing quickly enough.

The first five weeks slid along in a muggy summer haze. The heavy heat seemed to follow me wherever I went, from the grocery store to the clammy depths of the subway system. By the end of December, 2012 would be named the hottest year on record in the US. Despite the unpleasant weather, I filled my days with exploration. I visited community gardens around Manhattan, took a ferry to Governors Island, saw movies and went to free festivals. I made the mistake of trekking four miles to Chinatown on a 90-degree day, armed with only one bottle of water, and succumbed to the symptoms of heat exhaustion—headache, nausea, dizziness, and cramps. Luckily, spending the rest of the day in the darkness of our railroad-style apartment and guzzling ice water was all I needed to remedy my malaise.

But two weeks later, just after starting my classes on photography, videography, and radio reporting, a new array of symptoms struck. Sitting in a cool hall for a lecture, I noticed that my hands and arms were shaking. The tremors only grew more intense as I headed back to our Hell's Kitchen apartment, at which point dizziness and breathlessness were added to the mix. Had I become overheated again? This time, no amount of water or fans blowing in my face provided any relief. After several hours of waiting to feel some improvement, I walked to the nearest urgent care. The drugstore nurse speculated on the possibility of diabetes or a blood clot in my lung, but a few tests ruled both those out. With no answers, just instructions to go to the hospital if anything got worse, I was sent home. Several days later came a call with my lab results: something was wrong with my thyroid.

The first specialist I scheduled an appointment with denied me care because my insurance was out of network, a term I'd never heard before because there were so few doctors in my hometown. I was more careful scheduling the next one, but I still avoided learning anything about the thyroid. I'm not sure I could've told you where it was located until a white-haired endocrinologist pressed an ultrasound wand against my throat. Then, as I lay flat on my back, the ultrasound wand exerting a choking pressure on my neck, the seriousness of the problem became more apparent.

My thyroid appeared on-screen in grainy resolution, its shape vague. Only fleecy white static with a smattering of black spots. Those spots, the doctor explained, weren't supposed to be there. They represented the damage wrought by my immune system. The cells meant to defend my body from viruses and bacteria had mistaken native tissues as an invader.

The medical term for those empty patches is "black holes," and they appear with end-stage Hashimoto's thyroiditis, my disease.[26] In other words, the doctor explained, this attack on my thyroid gland had been occurring for some time. It wasn't a new phenomenon. My thyroid was so damaged that it could no longer regulate the amount of hormone it released into my bloodstream, so it alternated between a flood and a trickle. And those hormones affected my brain, heart, muscles, metabolism—and my ability to regulate my body temperature.

In space, the center of black holes approach absolute zero, a temperature so unfathomably cold that molecules in gas stop moving. But the event horizon around them, where radiation and matter collapse, is the opposite: hundreds of millions of degrees, the heat of friction from a whirlpool that nothing can escape.[27] Was my dying thyroid hot or cold? I didn't think to ask. I didn't even put together that my strange temperature sensitivity was related to this gland. My brain failed to make the connection between the autoimmune onslaught and the many odd symptoms I'd experienced over the past six years. I was too shocked.

The endocrinologist removed the slick wand from my throat and asked about my symptoms: had I experienced any hair loss, a hoarse voice, fatigue, or strange menstrual cycles?

I still didn't understand what the thyroid was meant to do, or how mine could function despite being riddled with holes. If a heart is punctured that many times, surely a person would die. How important could this butterfly gland in my neck be if I was still alive even as it was disintegrating? And at the same time, how much worse would I feel if it continued to be attacked? In that office, on the crinkly white paper of the exam table with ultrasound gel dampening the collar of my shirt, I could only tell the doctor that I felt shaky and anxious and not like myself.

The doctor said he would expect as much, given the state of my thyroid. Thyroid hormone replacement would make me feel better, and the generic medication was cheap. I remember him proclaiming that T4 therapy was a better treatment route than T4/T3 combination therapy, pointing to a black-and-white photocopied article pinned to the wall to support his case.[28] He also told me that some patients asked about Armour, a medication made from the thyroid gland of a pig, but that the synthetic version was better.

I didn't understand half of what he was talking about; it would be months before I learned to differentiate the various thyroid blood measurements, and years before I learned the history of thyroid disease and how it was once treated by giving people animal thyroid glands to ingest.[29] I also didn't understand this rhetorical move: he was heading off future questions I might bring him, denying me the possibility of being in charge of my treatment plan. He didn't explicitly say any of those things, of course; it was something I could only identify after many more encounters with doctors. At the moment of diagnosis, I wanted to trust the doctor. I wanted to believe he'd make me better, even though he did so little to prepare me for what might come next—such as developing other autoimmune diseases or having strange reactions to over-the-counter medications (like pseudoephedrine), which can interact with thyroid hormones.

The diagnosis provided a short-lived sense of relief. I had a treatment; I would improve. But alongside that hope was a mountain of inarticulable questions. Before now, illness seemed to exist on two poles: the inconvenient or the deadly. This privileged understanding of bodies and health is endlessly propagated by our culture, from the wellness industry to the ableism that conflates a person's worth with their abilities. There is no cure for most autoimmune diseases, only treatments to manage symptoms.

I was sick; I had been sick before I had the words for it; I would always be sick. Where did that leave me? In a place of impossibilities. I had a name for a disease, but no name for myself. I couldn't believe what was happening to my body, even with all the evidence accumulating in my medical files. I didn't have language to describe it and I didn't want to find that language. If I left everything unsaid, it wouldn't have to be fully real. Yes, I had to take medicine and go to appointments. But that was the end of it. If I looked at the impact of this disease on my body—if I contemplated where it might lead me in three years, five years, ten? No. Might as well see how long I could stare at the sun before going blind. Denial is just another name for self-protection.

HEAT WAVES, HORMONES, AND SELF-CONTROL

The simplified version of temperature on our planet is an interplay between the Sun, Earth's various surfaces, and gases in our atmosphere. Solar radiation is reflected or absorbed by the atmosphere and different types of terrain, with about 30 percent of the sun's radiation being reflected out to space. With more precision comes more complications. Oceans absorb CO_2, volcanoes belch it out, humans dig up fossilized carbon and burn it, adding 100 times more CO_2 every year than all active volcanoes combined.[1] The temperature rises, and is rising.[2]

A similar relationship between heat production and maintenance exists in our bodies: the hypothalamus and pituitary gland in the brain communicate with the thyroid gland, catalyzing it to release hormones that help regulate body temperature. The hormones are involved in metabolism (the chemical reactions in cells that provide the body with energy) as well as blood vessel dilation—which, when the body is not functioning properly, can make a person too hot or too cold, too sensitive to the changing temperatures around them.

Thyroid hormones also affect heart rate, skin and bone maintenance, fertility, breathing, digestion, and cognition. A gland only two inches long has a hand in so many organ systems that it can literally reshape the entire body. An especially pronounced example of this is congenital hypothyroidism (historically called "cretinism," a derogatory term).[3] In this disease, the thyroid either doesn't develop or doesn't produce hormones as it should. The result, if left untreated: the body doesn't grow at a normal pace, with children retaining a small stature and experiencing intellectual disability.

The transformative powers of thyroid hormones are equally dramatic among amphibians. Most frogs start their lives as wriggling tadpoles and only grow limbs when thyroid hormone levels rise.[4] Even more remarkable are the axolotls, a salamander with a smiley face and feathery gills sprouting behind their cheeks. They typically don't metamorphose because they live in a permanent state of hypothyroidism, with very low levels of thyroid hormone in their bloodstream.[5] Because they retain their gills, they can spend their entire lives in water, and their juvenile body traits don't prevent them from mating.

Scientists have used axolotls for the development of wound-healing products and to test detrimental impacts of certain toxic chemicals. They're also frequent subjects of lab experimentation in limb regeneration, with axolotls getting their limbs and tails severed to see how well they can repair damage.[6] To make their wounds even more applicable to human injuries, one experiment involved crushing their spinal cord to see how well they repaired the damage.[7]

But humans are affecting their life cycles in other ways as well. The ever-increasing abundance of CO_2 in the atmosphere means a heavier blanket wrapping around the planet, trapping more heat. And some amphibians have limited tolerance for temperature variations, owing in part to their need for water at certain stages of the life cycle, and for land at others.[8] Increased heat and longer droughts are causing serious problems for the amphibians. When temperatures grow too warm, common hourglass tree frogs have trouble advancing beyond the tadpole stage.[9] If it's hot enough, the frogs simply die off. This is also the case for axolotls, who can't tolerate high temperatures that other newts seem to manage without much trouble.[10]

I saw an axolotl in a zoo aquarium not long after being diagnosed with Hashimoto's. Its pale face and pink gills greeted me from behind clear glass; a nearby placard described its critically endangered status in the wild. Axolotls reproduce readily in captivity. The problem is that their wild habitats have become less and less hospitable. The high-altitude lakes around Mexico City where they evolved have been dried up or polluted throughout the twentieth century and into the twenty-first. The axolotls once had few natural predators; now common carp and tilapia released into these waterways prey on them.[11] Thanks to their popularity as a symbol of Mexico, plenty of locals are invested in their recovery, from beer brewers to farmers.[12] But even with such ecological restoration efforts, there's still the problem of temperature. Mexico City is getting warmer, like the rest of the world, and urbanization creates pockets of even higher temperatures. There are somewhere between fifty and a thousand mature wild axolotls.[13] What if they don't survive climate change, even after they've survived anything else? Their populations may all be confined to zoos and home aquariums, unwitting climate migrants who are allowed entry past policed borders because we find them cute and innocuous.

It's easy to feel an affinity with such a quirky amphibian. Though its permanent hypothyroidism is a natural state, unlike mine, which was the result of disease, I still had the sense that both our bodies were subject to forces beyond our control. Neither of us could handle the extremes of heat or cold. Both of us had to live in uncertain worlds, at the mercy of other humans. For the axolotl, this means receiving care from aquarium owners, scientists, and dedicated conservationists. For me, it meant relying on the judgment of doctors, experts I'd been trained to trust rather than question. We both lived on the borderland of medical science. The axolotl's physiological oddities, from hypothyroidism to limb regrowth, have made it a subject of endless fascination and experimentation. My autoimmune disease slotted me into the growing population of people who are chronically ill, the majority of them women.[14] Our existence was a source of questions, speculation, and experimentation. We no longer belonged fully to ourselves.

The new rhythms of my life included semimonthly blood tests and trips to see my endocrinologist for medication adjustments. Eventually, he said, I wouldn't have to come in so often. But in the first year, we'd likely have to adjust my medication dose fairly frequently. My endocrinologist still had a tendency to treat me as an unreliable narrator of my bodily experiences. Our bodies are so sensitive to thyroid hormones that medications are doled out in micrograms instead of the more common milligrams. An average dose of regular-strength Tylenol is 325 milligrams. My daily dose of thyroid hormone at the time was 100 micrograms—more than three thousand times smaller.

Yet it still felt like too much. When I returned for a follow-up appointment two months after my initial visit, I described a series of worsening symptoms. I'd be pacing in my apartment, trying to think my way through homework, but unable to write because my hands were shaking too badly. I was often tired but struggled to sleep. Panic attacks ripped through me on a regular basis, seemingly without any external trigger.

"I recommend seeing a psychiatrist," the doctor said. "You're probably stressed from grad school. There's nothing more I can do for you."

I left, fuming and dejected. Days later, as my symptoms grew worse, I trekked once more to an urgent care. When the results of my blood work came back, the doctor from the urgent care called me in a panic. "You're taking way too much thyroid hormone medication," he explained. "Your doctor never should have put you on such a high dose. This medication can kill you if you're not careful!" The urgent care doctor was horrified that I might experience a condition called "thyroid storm," which is when a drastic increase in thyroid hormones produces fever, delirium, irregular heartbeat, low blood pressure, coma, and even death.

This should've been my sign that it would be prudent to look for a new endocrinologist. Instead, I went back again, convinced he would apologize and take me more seriously the next time I complained of mood swings. By then I had done some research and knew that anxiety and depression are direct symptoms of over- and under-medication for Hashimoto's thyroiditis. What I had not yet learned was that women have historically faced skepticism and dismissal from medical professionals, and that such treatment continues to be common—with sometimes deadly consequences—to this day.[15]

As fall turned into winter, I tilted into the opposite end of thyroid symptoms. Gone was the surplus of energy, the anxiety, the insatiable hunger. In its place came depression, sometimes so heavy I could almost believe I'd buckle under the weight of the atmosphere, as if it were miles and miles of water drowning me. But when I tried to explain this to the endocrinologist, I was told the same thing again. Higher education, living with a partner, moving to a new city—all are stressful experiences. My thyroid couldn't be blamed for every little thing.

The summer of 2013, I began looking for a new endocrinologist. I'd put up with enough mistreatment and deserved a better relationship with someone who would be monitoring my health on a regular basis. My symptoms remained irregular and unpredictable. I needed help in managing the havoc wrought by my immune system.

An online review service led me to a highly rated endocrinologist in Brighton Beach, who specialized in diabetes but also treated thyroid disease. Visiting his office required a long subway ride to the edge of the borough, abutting Coney Island. The latter half of the ride brought the train aboveground, and I reveled in the cityscape stretching out to the Atlantic Ocean. I could almost believe myself no longer in the city, but having escaped to a beach resort town, the salty sea breeze lifting the summer stench of the sewers and human sweat. That is, until I entered the overly air-conditioned doctor's office and was met by a crowded waiting room.

Each visit to this new doctor brought new frustrations, even humiliations. Long wait times meant I couldn't predict when I might be back to the office for my first post–grad school job. The endocrinologist was by turns surly and congenial, his moods impossible to predict. He required that I have testing for diabetes, despite showing no signs of it. When I was made to wait an extra hour beyond my appointment time for a fasting blood draw, I became so lightheaded that I stole a bite of food, which disrupted my test results and led to the doctor berating me. A second round of testing revealed that I did not, in fact, have any signs of developing diabetes. My hypothyroidism instead predisposed me to hypoglycemia, or low blood sugar, which could explain why I'd become so lightheaded and shaky when required to fast so long for the blood test.

This doctor, like the last one, was unable to offer advice on how to manage the jarring swings between hypo- and hyperthyroidism that accompanied my illness and treatment. I resigned myself to another stultifying summer and determined to make the best of it by becoming even more active, joining a rock-climbing gym, doing yoga classes, and lifting weights. I even became, for the first time in my life, someone who ran regularly. Summer saw me doing 5Ks around Prospect Park in Brooklyn, striving for a better time. When my partner signed up for a Seattle marathon, I decided to attempt the half. Health, I still believed, was something a person could earn through the correct choices and proper behavior.

It was the epitome of a neoliberal mindset, a political-economic ideology that states economies and societies should be organized along principles of the free market, which means decreased government regulation and increased free trade. Becoming the status quo in the 1980s under American president Ronald Reagan and British prime minister Margaret Thatcher, the ideology expanded from a political and economic alignment to a series of cultural beliefs and practices.[16] As geography scholar Matthew Sparke writes, people are "routinely told that their health is simply their own responsibility, a form of resilience that will only endure if they invest in it with the same individualistic and entrepreneurial prudence that is the trademark of personalized neoliberalism more generally."[17]

How many humans spend their lives chasing this mirage, seeking control over every aspect of health, bodily function, and appearance? Buying and consuming spurious supplements, following extreme diets, even trying to manipulate the way they respond to temperature. Take "The Iceman," a Dutch extreme athlete who has climbed parts of Mount Everest in shorts and set records for spending nearly two hours in an ice bath. He coined the eponymous "Wim Hof Method," a practice of breathing and cold therapy that he claims has health benefits such as increased energy and a stronger immune system. There are also wellness influencers who tout the health benefits of cold plunging despite a lack of sufficient research to justify such claims (and even some evidence that sudden immersion can be dangerous). And on the opposite end of the temperature spectrum is hot yoga, with its purported benefits to flexibility and cardiovascular health (which, again, comes with risks).

An emphasis on personal responsibility is also apparent in messaging about how to reduce greenhouse gas emissions. Oil and gas giant BP introduced the carbon footprint calculator, letting individuals measure how everything from their diet to travel contributes to greenhouse gases entering the atmosphere. This, of course, obscures the reality that just twenty fossil fuel companies (including BP) are responsible for more than one-third of all emissions since 1965—and that they've known about the dangerous climate effects of fossil fuels for more than fifty years.[18] In 1968, scientists commissioned by the American Petroleum Institute warned that if fossil fuel production kept growing, "significant temperature changes are almost certain to occur by the year 2000. . . . There seems to be no doubt that the damage to our environment could be severe."[19] Companies like BP, Chevron, and ExxonMobil knew about the alarming effects of fossil fuel production, but rather than share this information with the public, these companies denied and obfuscated.[20] In a 1988 memo, an Exxon spokesperson noted the company's goal was to "emphasize the uncertainty in scientific conclusions regarding the potential enhanced Greenhouse effect."[21]

As more scientific evidence accumulated through the end of the twentieth century, scientists began sounding the alarm publicly about the dangers of climate change, including in reports by the Intergovernmental Panel on Climate Change. Its 2007 assessment stated that the science showing human fossil fuel use was behind climate was now unequivocal.[22] Yet the George W. Bush administration continued to deny the science. "Suppression of reports and behind-the-scenes falsification of documents were accompanied by an official approach that the problem 'needed further study,'" writes environmental sociologist Kari Marie Norgaard.[23] By sowing doubt about the science, Republicans and fossil fuel companies created a climate of denial that continues to this day. Almost 15 percent of Americans do not believe in climate change.[24] Nearly 30 percent believe that warming is caused by natural patterns in the environment, not human activity.[25] Meanwhile, fossil fuel companies collectively make trillions of dollars annually, and 2024 was the first year in which the average global temperature was more than 1.5°C higher than it was in the era before fossil fuels.[26]

Neoliberalism and the policies it engenders—austerity, deregulation, globalization, etc.—create inequality.[27] That insecurity is a feature, forcing individuals to be more competitive, to strive for the reward that is always just out of reach. It hurts far more people than it helps.[28] But I don't believe that passively accepting the status quo is the answer. When individuals feel completely disempowered to do anything against massive corporations—when we relinquish all responsibility—this helplessness can turn into a "denial of self-involvement" that allows people to avoid personal responsibility because higher authorities or a large number of people are the source of the problem, so an individual's actions become negligible.[29]

We need to balance the knowledge that we have autonomy with the understanding of our small role in a huge web of actions and behaviors. As social psychologists Susan Opotow and Leah Weiss write, "First, we are all victims in that we are recipients of pollution generated by others. Second, we are all violators in that we create pollution that has an impact on others. Third, we all need to work at ongoing, constructive problem solving and dialogue."[30]

I hadn't yet learned this lesson in cooperation and care in the early years of my thyroid disease; embracing that philosophy would require relinquishing too much control over the trajectory of my body. But where I couldn't yet see the interconnectedness of health and bodies, mine and others, I was beginning to see it in the environment around me.

The basic difference between climate and weather comes down to time, place, and patterns. The temperature may swing from 15°F to 50°F on a blustery day in March, snow warming to rain. The next day might be freezing again, the sidewalks slicked with ice, grass trapped in a shimmer of frost. All this is weather: local ripples of cold and warm fronts; moisture from the soil and the trees interacting with the air; where we find ourselves on Earth's annual precession around the sun. The isolated phenomenon of a heat wave is weather; its intensity, when it falls in the calendar year, and its duration are all components of climate—the longer-term, larger-scale patterns. This is part of what makes climate so hard to comprehend on a human scale.

Because past climatic changes unfolded much more slowly, and the human lifespan is generally short of a century, it would've been hard to perceive such changes in a lifetime. At least, so we thought until recently. New research on "abrupt climate change" looks at wild fluctuations that happened in the span of decades or even years.[31] One such event is called the Younger Dryas, named for a white flower with a spray of yellow stamen at its center. This Arctic perennial suddenly became widespread across Europe about 12,800 years ago, when a glacial period interrupted what had been a warming cycle.[32] In only a decade, temperatures dropped as much as 18°F in some regions of the Northern Hemisphere.

Researchers have hypothesized that this abrupt change may have been caused by a huge influx of freshwater into the Atlantic from melting ice sheets, disrupting ocean currents to such an extent that the global climate shifted very abruptly. Others have proposed that a comet collided with the planet, initiating widespread wildfires that filled the skies with soot, contributing to cooling.[33] This wouldn't have been anything like the asteroid that caused the extinction of the dinosaurs, but the Younger Dryas event still took a heavy toll on creatures alive at the time. Around 80 percent of megafauna like ground sloths, mammoths, and *Camelops* died out in North America.[34] This extinction was previously argued to have been caused by human overhunting but may be related to this sudden temperature swing.[35]

Such a sudden shift seems like it should belong only to sci-fi films like *The Day After Tomorrow*. Even after learning more about climate change as a university student, and seeing its effects as the years went by, I still felt a sense of predictability. Winter in the higher latitudes of the Northern Hemisphere brings low temperatures and snow. Buds open in the spring, trees wave verdant limbs in the summer, leaves glow red and orange in the fall. These were the normal patterns of the seasons, creating constancy and confidence about how the weather behaved from one month to the next.

But as scientists have been learning, abrupt change is possible. If we look to the organisms with whom we share the planet, it should be obvious that something dramatic is happening: some 28 percent of species are currently threatened, and if we want to avoid a mass extinction, humans will need to keep warming to less than 2.5°C by 2100.[36]

Collectively, we know human activity and fossil fuel burning is causing global warming. Medically, I know my immune system has spent more than half my life attacking my thyroid and making me sick. But even with an abundance of knowledge and ongoing research, predictions are challenging. There's always the possibility that something dramatic will occur, fundamentally changing life.

By the time I reached my thirties, my thyroid disease felt familiar and predictable. Many more illnesses emerged over that period, creating their own havoc, new weather patterns. We'll get to those soon enough. But my thyroid—riven with holes, unable to produce enough hormone—I thought I'd come to understand it. Bimonthly blood draws had transitioned to biannual, as my hormone levels stabilized. Occasionally I needed to change my medication dose, but the regular swings from hypo- to hyperthyroidism I experienced the first two years after diagnosis were gone. We'd settled into a more amical relationship, my thyroid and me.

Until, very suddenly, everything changed. Again.

I didn't notice the weight loss at first because I wasn't in the habit of weighing myself. But after someone pointed out that I appeared to be losing weight, I started checking weekly. I watched the scale's number slip down with a frightening speed. I hadn't changed my diet or exercise or anything else. That was the first sign.

Then there was the difficulty falling asleep, the feeling of my heart pounding so hard that it echoed in my ears, as if I were in a night club. I couldn't drown it out. Next came the extreme heat intolerance. I was suffocating even on the cooler days of August, walking around the house with my shirt rolled up to my bra, placing ice packs around my neck, sweating into the eye mask I wore to bed at night and waking with cold, clammy cloth pressed into my face. Finally, one weekend morning I got up to make cornmeal pudding, a Puerto Rican porridge from my mom's childhood, and I nearly collapsed over the hot stove. I moved to a room with a ceiling fan on high and drank ice water and ate sugary snacks in case my blood sugar was low. My doctor wrote a referral to an endocrinologist.

The office was on the third floor of a hospital with a nurses' station in the center of a long corridor. One of the nurses led me to the examination room, which was strangely noisy, as if a helicopter were landing or taking off above us. When she weighed me, I noted that I'd lost over twenty pounds in the past few months. When she took my pulse and blood pressure, she asked if I was nervous; my heart rate was 124 beats per minute. I wasn't nervous so much as desperate and confused. This wasn't supposed to happen.

The endocrinologist came in, a woman who had me hold out my hands (the fingers vibrated visibly, a symptom that made it increasingly hard for me to write because my pen lines turned to scribbles) and then palpated my neck. She announced I'd developed a little goiter. She asked about my other symptoms and I described the heat intolerance, the trouble sleeping. Worst of all was the muscle weakness. Walking downstairs was a struggle, as I constantly felt on the verge of collapsing. Walking to the park was momentarily out of the question, both because I'd grow faint from the heat and because I didn't trust my body to carry me that far.

Almost twelve years exactly since the very first endocrinologist prescribed levothyroxine and told me I'd be taking the thyroid hormone replacement for the rest of my life, this new endocrinologist told me I'd need to stay off levothyroxine and prescribed me an anti-thyroid drug. She thought I was experiencing Hashitoxicosis, a hyperthyroid phase of Hashimoto's, but she needed to check on my thyroid with an ultrasound. It wasn't clear what might be happening, because Hashitoxicosis usually happens much earlier in the progression of the disease.

The very next day, I started propylthiouracil, a drug that's also been used for research into taste sensitivity. I found it so bitter that if I didn't throw it to the back of my mouth to swallow, my gag reflex kicked in and made the pills nearly impossible to get down. I had to take them three times a day. And every time I gulped down those noxious, awful white pills, I was reminded that whatever I'd thought I understood about my thyroid had been wrong. That in anticipating the continuation of a known pattern, I'd been lulled back into denial. I'd let myself believe, again, that I was still in control.

DEATH VALLEY AND DEVILS HOLE

Death Valley, California, has the hottest recorded air temperature of any place on Earth; it reached 134°F (57°C) at Furnace Creek in 1913.[1] In known history, it has twice gone an entire year without any rain; the average high temperature is above 100°F for five months of the year. Today it's a National Park; red signs at trailheads admonish visitors that "Heat Kills!" Others warn of "Extreme Heat Danger." Its geology and geography both play roles in giving the valley its superlative status. Although there are over a thousand plant species found around the park, the low-elevation salt pan prevents any growth, and other parts of the valley have limited foliage, so the sun largely beats down on bare rocks and sand.[2] Because the valley basin is 282 feet below sea level and hemmed in by high mountains, the hot air is trapped in a small bubble. Atmosphere and topography together create some of the most extreme thermal conditions that exist on our planet, and they're getting more intense: the summer of 2024 was the hottest on record for Death Valley, with only seven days in July when the park temperature failed to reach 120°F.[3]

But some 170 years ago, when the earliest gold rush settlers were traveling overland from the east, the desert had yet to earn its ominous name or reputation. It was, in fact, home to the Timbisha Shoshone, a tribe that for at least one thousand years had lived on and understood the land, which they called *Tüpippüh*.[4] They hunted wildlife, like bighorn sheep and mule deer, and relied on honey mesquite plants for their pods and pinyon pines for their seeds.[5] Family groups moved around the landscape, leaving the hot valley floor in the summer for seasonal camps in the cool mountains. As Timbisha elder Pauline Esteves would write more than a century into the colonization and displacement process, "The term 'Death Valley' is unfortunate. . . . This is a place about life."[6]

Even though Euro-American settlers certainly came across these groups of Timbisha Shoshone, that didn't prevent them from characterizing the landscape as empty. The government denied Indigenous people's connection to the land and allowed prospectors and speculators to make use of water and other resources without regard to the Indigenous people.[7] In 1933, President Hoover established Death Valley National Monument without recognizing tribal rights to the land.[8]

In 1849, around the start of the California gold rush, fur trader William Lewis Manly was among a group of settlers in a wagon train whose food supply dwindled while attempting to cross the desert in winter. He and another young man were sent ahead to scout for a settlement and food while the other members of the party waited at a spring. As they trekked through canyons and up mountains, they came across the body of a traveler from a different wagon train.

"Our mouths became so dry we had to put a bullet or small smooth stone in and chew it and turn it around with the tongue to induce a flow of saliva. If we saw a spear of green grass on the north side of a rock, it was quickly pulled and eaten to obtain the little moisture it contained," Manly wrote in his autobiography.[9]

Though Manly and his partner had provisions, including meat, their mouths were too dry to swallow the food. Their one saving grace after multiple days without water was walking along a mountain ridge at night and discovering a frozen puddle of water; by sunrise, it would've melted into the dry earth. At many moments, the only thing keeping the two men moving forward was the thought of the party they'd left behind, which included women and children. "No one who has ever felt the extreme of thirst can imagine the distress, the dispair [*sic*] which it brings," Manly wrote. "I can find no words, no way to express it so others can understand."[10]

Dehydration was an obvious concern for the settlers; a human can only survive without water for around three days. But the heat itself exerts tremendous pressure on the body. Sweat comes first, a mixed blessing as a cooling mechanism since it can contribute to dehydration. Blood vessels nearest the surface of the skin widen to assist with cooling. If the heat causes a person's body temperature to rise, their heart rate increases by seven beats per minute for every one degree Celsius of temperature increase.[11] Small blisters can form around the groin and elbows and armpits; cramps ripple through muscles. All these physical sensations can exacerbate mental health problems like anxiety, depression, and bipolar disorder.[12] They also tend to make people quick to anger. Violent crime increases with the temperature, though heat isn't the only factor to consider: researchers have found that a higher amount of vegetation is associated with much lower rates of outdoor crime, possibly because greenery in urban spaces reduces the temperature.[13]

If heat exhaustion proceeds to heatstroke, the body cooks. Kidney failure often comes first, which can be followed by blood poisoning and the disintegration of the gut lining. Heat has been the leading cause of weather-related death in the US for decades, killing more people each year than hurricanes, tornadoes, floods, or cold.[14] Some research suggests heat-related deaths in the US might exceed 1,300 people each year.[15] Even if we manage to limit global warming to only 2°C—a goal that has a 50 percent chance of being met only if we achieve net-zero emissions by the 2050s—people living in the tropics will be exposed to dangerous temperatures *most days of the year* by 2100.[16] Heat waves like the one that smothered the Pacific Northwest in 2021, killing nearly one thousand people, are predicted to become regular annual events.[17]

Despite their extreme suffering, Manly and his partner found shelter, food, and water after days of walking. They ultimately rescued the members of their wagon train who were left behind. The men were fortunate in being young and fit in advance of their trials in the desert. Their party gave Death Valley its name, but their survival is what allowed them to do so.

The heat may exert an equal force on all bodies, but not all bodies have the same physiological response. The very old and the very young are at higher risk of heat-related illnesses, as are those with any form of dementia. Certain medications affect thermoregulation, sweating, and electrolyte balance, predisposing people to heat exhaustion. Then there are diseases that directly affect how the body generates and interacts with heat—some of them so subtle that it may be years after their onset that the effects are detected.

My hyperthyroid symptoms improved slightly after a few weeks of taking the foul anti-thyroid medication, propylthiouracil, and a beta blocker to lower my heart rate. I was still walking around with my shirt half off, still shaking. In advance of my ultrasound appointment with the endocrinologist, I feared she might discover a mass of nodules growing across my thyroid, or some cancerous tumor. The day of the appointment, I prepared myself for that outcome.

This time the ultrasound wand pressing against my throat was almost painful. My thyroid had felt sore for weeks, occasionally sending discomfort radiating up my neck. The doctor didn't show me the screen like the first endocrinologist did all those years ago, nor did she offer any commentary during the exam. Instead she waited till she finished shining sound waves on my thyroid gland and passed me a tissue to wipe gel off my neck.

"There are no nodules or holes," she said at last. "It is a solid thyroid that is highly vascularized. You have Graves' disease." What she didn't say (but I discovered later in my own research) is that this classic appearance of Graves' is sometimes called a "thyroid inferno."[18] How had I gone from black holes to an out-of-control fire in the span of a decade?

I sputtered out questions, trying to understand the transition from one disease to its opposite. The doctor shrugged; the immune system changed its attack and we can't explain why. She was more interested in pursuing the best treatment, leaving me to make sense of the metaphysics of this new disease process. Everything I read suggested it's rare to switch from Hashimoto's to Graves', but nothing I found explains why it happens.[19]

I felt a bit like the NASA scientist whose name has appeared in recent articles grappling with the extreme heat of 2023 and 2024. In an article for *Nature*, climatologist Gavin Schmidt writes, "It's humbling, and a bit worrying, to admit that no year has confounded climate scientists' predictive capabilities more than 2023 has."[20] In an interview with journalist Elizabeth Kolbert, he was even more blunt. "People started using adjectives that scientists don't generally tend to use [for the 2023 heat]," Schmidt said. "Our eyebrows at this point were rolling over the top of our heads."[21] In other words, something is wrong with the models, hindering their ability to forecast what comes next.

The scientist's shock and confusion mirrored my own when I faced an unexpected change in my thyroid disease. No doctor ever warned me that this might happen. I hadn't known it was a possibility, even though it always was. We all want warnings, predictions, ways to penetrate the future, because being prepared offers a sense of control. But even if some endocrinologist early in my time with thyroid disease had said, "Someday you might switch to hyperthyroidism and Graves' disease," that knowledge wouldn't have saved me from the discomfort of hyperthyroid symptoms.

The unexpected rapidity of climate change doesn't offer any comfort, but the knowledge that even the best researchers can be stumped is a useful reminder: we're nowhere near an endpoint in understanding the weather or the human body. After my endocrinologist concluded the ultrasound exam and explained future options—using radioactive iodine to destroy part of my thyroid gland; surgery to remove it entirely—I tried to make peace with not knowing the "why." Our bodies aren't under our control nearly so much as we'd like to believe. Rather than deny this new change as an impossibility, I'd have to accept that sometimes, the most unexpected things happen.

When Manly and his party passed through the Mojave Desert, toward the terrain that would be renamed "Death Valley" thanks to their experiences, several members of another wagon group that met with Manly's found a narrow strip of water surrounded by rocky cave walls. German immigrant Louis Nusbaumer wrote in his journal, "The temperature of the water is about 24–26° and the saline cavity itself presents a magical appearance."[22] Did he or any others notice the little blue fish flitting around the pool? They make no mention of the aquatic life. In any case, they couldn't possibly have imagined the strange future of this magical spring: that it would become a bathing site for miners, then have its water levels lowered by people pumping wells for the irrigation of a ranch, and then ignite a Supreme Court water rights case pitting fish against farmers. Nusbaumer and his friend didn't even have a name for the little pool. But today, we call it Devils Hole.

My partner and I crossed into the desert in an air-conditioned SUV rather than a wagon, our backseat loaded with a gallon jug of water, plus water bottles, plus electrolyte tablets and powders. We carried jerky and salted pistachios and chips and cookies in hopes of staving off the sodium imbalances that can happen with excessive sweating. We followed a dusty road through Ash Meadows National Wildlife Refuge till we came to a rolling hill surrounded by barbed-wire fences and security cameras. Home of the Devils Hole pupfish, *Cyprinodon diabolis*, a critically endangered fish living in the smallest habitat of any vertebrate.[23] The only people who swim in this pool now are the specially trained divers wearing gear that is only used for this particular site. They enter the 92°F water twice a year to count fish.

A crowd just shy of two dozen people gathered near the fenced entrance for the late-September event, some pulling on wetsuits while others erected shade canopies. Biologists from the National Park Service, Nevada Department of Wildlife, Ash Meadows National Wildlife Refuge, and US Fish and Wildlife Service—plus some volunteer divers and a couple of journalists. Kevin Wilson, a supervisor biologist for the National Park Service and lead organizer, started the day with the temperature. "We're projecting over 100 degrees today," he said as a bighorn sheep meandered down the hill alongside the fence. "Typically we're in the nice eighties or so, maybe low nineties, so we really have to pay attention to staying hydrated. My motto is 'Clear, copious and often,' and that's when we should be peeing."

"If you are feeling remotely ill, I just want to point out quickly what heat stress feels like," added Amy Fowler, safety manager for Death Valley National Park. She'd brought multiple hats, coolers full of ice water, and misting fans. "If you are starting to get remotely like a headache, you're probably starting to have heat stress. If you are starting to feel a little nauseous, that's another sign that your body is trying to get rid of heat, so it doesn't want to hold on to food in your stomach," Fowler said. "Be mindful of what your body is doing and listen to it—don't try to power through."

I'd been taking anti-thyroid medication for five weeks at that point and no longer experienced extreme heat intolerance, but I wasn't taking any chances. I'd purchased fast-drying lightweight clothes for the outing, as well as a cooling necktie. It wasn't yet 8 a.m. and the heat was already intense. Even knowing the risk of visiting the desert after months of hyperthyroidism, I was desperate to see these fish. I wanted to understand how anything could survive in an environment that seemed inimical to life.

That the pupfish even made it to Devils Hole is a fluke of geology. Around sixty thousand years ago, a rock ceiling collapsed into a crevice that opens into the Amargosa Valley aquifer. Not only did those chunks of stone create an opening to the sky that allowed a new ecosystem to form; they also got stuck at various depths in the water. At the water's shallowest, rocks created a ledge that receives sunlight in the spring and summer, thus creating a food chain that begins with photosynthesis. In the winter, the fish rely on whatever comes from outside: bugs, leaves, dead birds, owl pellets.[24] It's this tiny shallow shelf that gives the pupfish virtually all the requirements for continuing their population. It's where they forage, it's where they spawn. Even though the temperature on the shelf fluctuates dramatically, the warmth that builds up on hot days barely spreads. This strange shelf allows the fish to live in an environment that hovers around 92°F year-round, with about 2 milligrams of dissolved oxygen per liter.

As fish biologist Mike Schwemm explained while divers glided through the cave, "This is not a good environment for fish. They barely survive. The temperature is right up knocking at the door of what's too hot, oxygen is right at that limit, they're barely hanging on. And that's what makes it really interesting and such a novelty to preserve, in that it is something quite astonishing, and you would think that under these conditions over a long period of time that you would have an extinction event naturally."

Yet somehow, *Cyprinodon diabolis* have lived in this tiny pool of water for a long time—though the timing and circumstances of their arrival is still debated.[25] They may have been there for hundreds of years, or tens of thousands of years. At some point they were isolated from other species of pupfish in the region, whether because a shallow lake dried up, or because they were transported by something else, and they had to adapt. Imagine being suddenly transported to a world similar to the one you've always known, but with subtle differences. It's harder to breathe, harder to feed. Maybe it's not so bad in the first years, and each new generation adapts just a little more than the last one. It's not comfortable, but it's survivable. And then the water level suddenly drops, and the feeding grounds are gone. The things that made this hole in the rocks livable are almost entirely removed.

This is what happened shortly after the pupfish found their way onto the Endangered Species Preservation Act in 1967. Cattle ranchers pumped water from the same aquifer that feeds Devils Hole, using it to irrigate their property. They pumped to such an extent that the water level in Devils Hole dropped multiple feet, drying out a significant portion of the shallow shelf.[26] While litigation stretched on for years, the Supreme Court ultimately ruled in favor of the fish in 1976, since Devils Hole was a noncontiguous part of Death Valley National Monument and had prior rights to the water.

But the water level never recovered to its pre-pumping level, and since 2000, the entire pupfish population has dropped below forty individuals *twice*. The fish are so inbred that geneticists believe it's the equivalent of five to six generations of brother-sister breeding. At a certain point the fish biologists agreed to start feeding them supplemental food and adding cover packets to the water—essentially balls of sticks and twigs from the surrounding area.

"They would put those in so the larvae would have some places to hide from the adults, because they're just awful parents. Their favorite snack are baby pupfish and pupfish eggs," said Olin Feuerbacher, a biologist with US Fish and Wildlife who manages the only captive population of the fish. "The strategy may work when you have a few thousand in there, but when you're down to thirty-five fish, you're eating your own kids." He imagined the parents calling a baby pupfish over, joking, "'C'mere, Billy.'"[27]

The work done by Feuerbacher and his colleagues has created something of a buffer for the wild population. They've filled a hundred-thousand-gallon refuge tank with Devils Hole pupfish, meticulously learning how to raise them from eggs, then getting them to reproduce on their own. It hasn't been an easy process. When Feuerbacher took me to see the refuge tank, it was nothing like visiting an aquarium at a zoo. It was industrial, with tools hanging on the wall and steel railings around the edge of the platform, with a grated walkway a few feet down hovering above the water. A scuzz of algae covered the water, which extends into darkness, out to the parking lot. Like their brethren in Devils Hole, the pupfish here were given only a meager sliver of sunlight. They're also dosed with antibiotics, antifungals, and antiparasitics as eggs, which kills off a lot of the diseases they carry in the wild, including the fish version of tuberculosis. The water temperature is just a touch cooler, and the dissolved oxygen is a tiny bit higher, meaning they have it easier than the wild population.

But not by much. Their parents will still try to eat them. When they are larval fish, predaceous diving beetles tear them apart. They're 25 percent *more* inbred than the wild ones, which has led to an increased incidence of lethal heart defects. They never develop pelvic fins (used by other pupfish species to make rapid turns) because the high temperatures suppress their thyroid hormones, preventing the fins' development. As Feuerbacher told me, "Unlike most pupfish species that can adapt to very extreme environments at least for short periods, these, you basically look at them wrong and they're like, 'Well, I didn't like that. I think I'm gonna die now.'" Which is why it took more than fifty years of trying just to establish this refuge population. And why divers have been going down a hundred feet into Devils Hole multiple times a year for decades to count the whole pupfish population.

Despite the unusual heat the day of the fall count, the majority of Devils Hole was still in shade as we clambered down the uneven rocks. Someone warned about rock nettle (also known as desert stingbush, a name I instantly understood when a barb burrowed into my skin); another pointed out the abundance of aggressive bees flying over the water, with one occasionally chasing people. I watched the divers disappear underwater, then slowly come back to the surface, presaged by a wave of bubbles and the blue glow of their flashlights. At the surface, biologists leaned over the metal railing to count those visible around the shelf. All around, people talked about recent papers, and their own experiences diving, and other strange organisms they'd studied, like the Texas blind salamander. There was jockeying and joking between members of different departments as people volunteered for the surface count. It felt like a social gathering, a strange sort of party where the only drinking game was regular reminders to sip water and the main entertainment was tiny blue fish.

At the end of the second dive, once everyone was out of the water, I climbed down to the walkway above the shallow shelf. The bees were even louder here (one person had already been stung), but they seemed more interested in gathering mud and water than investigating a human. I crouched down above the open aquifer, trying to catch a glimpse of a fish. Shadowed by the rocks, I couldn't make out anything. I crawled from one corner of the platform to another, caught a glimpse of movement, felt a burst of admiration for the surface counters who'd spotted so many. It was hot even down here, and I was starting to feel the slightest touch of headache.

I thought about what Feuerbacher told me when we visited the refuge, how he feels humans have a responsibility for protecting these fish because we're the ones who screwed up their environment. But in addition to that, there's so much we still don't know about them. We don't understand what triggers spawning, or how they've managed to avoid extinction, or how their microbiomes affect their ability to survive in this environment. "If we lose the species, we lose the ability to learn from it," Feuerbacher said. "It might just be a really cute fish, but there also might be some really important things that we can learn from it. And if it's gone, we don't get to find out."

The Devils Hole pupfish are stressed and struggling, and they're also the canary in the coal mine. They're living at the very edge of what they can survive, and climate change is threatening to make their habitat completely inhospitable. Whatever happens to them will eventually affect the other pupfish species in the region, as all the waters warm. When the United Nations released a 2019 report stating that around one million plant and animal species are at risk of extinction, some within decades, it could easily have used the Devils Hole pupfish on the publication's cover.[28]

The Earth is indisputably getting warmer. The Devils Hole pupfish live in a precarious environment. Only a few degrees warmer, and they won't be able to survive—and then where will the refuge population go? Where does anyone go when conditions are too extreme for life?

In his book *After Sustainability*, philosopher John Foster describes three distinct types of denial that people tend to exhibit regarding the climate crisis.[29] First is the most straightforward: *literal denial*. This is the rejection of science, evidence, and the basic facts that have been known since Arrhenius first calculated the atmospheric effects of carbon dioxide in 1896. The second category is *interpretative denial*, when people accept some of the facts but reject their meaning. Climate change is happening, they say, but there's nothing we can do about it. This cynical perspective may be the safest strategy as far as preserving one's physical and mental energy. If the apocalypse is assured, why not live it up while the living's good?

Last of all comes *implicative denial*, the most insidious of the three. In this case people accept the science; they understand that climate change matters deeply and that we can do things to prevent the very worst outcome; but they stop short of fully engaging with the painful reality. *I still have to live my life*, goes the refrain. *What's one more plastic water bottle purchased from a drugstore on a hot day when the ocean is already swirling with garbage vortexes?* If we absorbed the magnitude of the calamity, the logical response would be to do absolutely everything in our power to keep it from unfolding. No more polyester clothing produced from petroleum, no more eating meat, no more disposable baby diapers. We would be in the streets daily, protesting this system of endless extraction. We would circulate lists of goods to eschew, of activities to avoid; we would experiment with new ways of living because this form of capitalism is killing our world. Accepting that reality is brutal, maybe impossible, but also necessary.

Psychoanalyst Sally Weintrobe frames denial in the face of climate change in a comparable way, but also highlights the role of neoliberalism, saying it "recruits people to participate in an immoral project, which is to live daily life in a way that collectively causes huge environmental and social damage and is unsustainable."[30] But humans aren't inherently selfish and unfeeling, Weintrobe adds, so neoliberalism has to delude us into engaging in "uncare"—we ignore social responsibility and the moral implications of our earth-destroying behavior.

Foster's implicative denial and Weintrobe's "uncare" are things I understand, because too often they dictate how I live. Much as I want to treat climate change as an existential crisis, life intrudes. There are bills to pay and dishes to wash and words to write. There are books I haven't read that I want to buy, non-locally grown food that I'm going to eat, toiletries I need that come in single-use plastic containers. I can boycott Amazon, remember my reusable grocery bags when I go to the store, avoid meat most days of the week. But I still need to drive to my doctors' appointments because Chicago's public transit system isn't robust enough to easily get me there. My medications are provided in plastic orange containers, some shipped over who knows how long a distance. It is deeply painful to think of all the ways my existence directly harms other living organisms, and yet, I insist on my own right to life and my own needs being met.

When I think of implicative denial, I think about how easy it would be to let the Devils Hole pupfish die out. These fish are balanced on a knife's edge between survival and extinction. A few bad years for their reproduction and they could easily disappear. The scientists all know this; they are not denying the precarity of the fish's situation or ignoring climate change. It is precisely in acknowledging the direness of the situation that they've been able to muster resources to protect the population. Everyone has accepted the reality and they're doing everything in their power to keep these weirdly resilient, ridiculously inbred little fish alive. That task has required creativity and ingenuity and compromise and disagreement. The scientists have made backup plans and backup plans for the backup plans. They've approached the survival of the pupfish with care and curiosity.

At its most powerful, denial obliterates creativity and curiosity. It narrows our vision to a pinhole, so that we can only see what is immediately in front of us and therefore believe that this way of living is the only possible option. Why does every household need a refrigerator? A dishwasher? An oven, a stove? We could have a community food center on every block, sharing resources and chores. Why do lawns require mowing? We could share our yards with insects and birds and other wildlife of our region. I don't know if these ideas are feasible, or what else could be done to change our patterns of uncare and resource extraction. But I know that denial doesn't only prevent us from feeling guilty about our current behavior; it also prevents us from imagining something better.

We engage in denial because it serves a useful purpose, protecting us from all kinds of unpleasant feelings. Denying our impact on the planet saves us from feeling moral injury; denying the death of a loved one staves off grief; denying our mortality keeps us from living in constant fear of what might happen to injure or destroy our bodies.

In the moments I've hidden behind denial—not only when I first got my thyroid diagnosis but at various other points in my experience with chronic illness and climate change—I've done so for very specific reasons. Often, it's been because taking a clear-eyed look at how my life will change is terrifying. I wanted to hide from the possibility of death and loss for as long as possible. But when your body and your planet are undergoing dramatic transformations, the fear will eventually find you. Species around the planet are dying off. We're not just living in an era of mass extinction; we're creating it. This horrifies me. But maybe I can't help being preoccupied with my own well-being. And for me, there is no terror more primal than the possibility of sudden death.

PART TWO

STORMS—HEART—FEAR

THE MEANING OF A STORM

Storms have haunted my dreams for as long as I can remember. Sometimes in these nighttime phantasms, twisters spin down from a clear sky and I sprint for a ditch or drainage tunnel. Sometimes I hear the sirens and huddle in a windowless bathroom while the walls shudder. Sometimes there's no tornado, only a sky going black, and hail so large it shatters the windows of the car I'm driving, then pummels me till the world is nothing but white static. Whenever I wake from these nightmares, my skin seems to sizzle with painful electricity, as though I'm still carrying the storm inside me. I want them to stop, and I also want to understand them. It's hard to live through different versions of the same dream and not wonder if it's a prediction.

The oldest written record of dream interpretation dates to 2120 BCE, carved into clay by a Sumerian, but I can't help believing the human fascination with dreams stretches much farther back.[1] Some thirty-six thousand years ago, ancient humans crept into the darkness of a limestone cave, the dancing light of torches making shaky shadows on the walls.[2] They sketched sprawling herds of megafauna—aurochs and horses and rhinos and lions. More than a thousand drawings, in red and black and white, documenting creatures that no longer exist, left behind in a country we now call France. The bones and pawprints of giant cave bears carpet the cave floor. Did the artists ever encounter those bears while going about their work? Did they dream of death by carnivores, by stampede, by a hunting injury that goes foul? Why was it so important to them to trace re-creations of these animals onto the walls of deep caves?

Similarly intricate engravings of animals grace the enormous stone pillars of Göbekli Tepe, dating to around 11,500 years ago.[3] The stone undulates with snakes and spiders and scorpions and wild boars.[4] Perhaps those carvings were invocations, using the terror of monsters to spur bravery. Or it could be another example of the many things that can undo a life. Whatever the explanation, our distant ancestors seemed ensorcelled by death, the need to make meaning from the creatures that could kill them. To pin down disaster before it occurred.

The only certain proof of humanity's longtime fascination with dreams is the scant texts on oneirology—the science of dreams—from Antiquity. Greek diviner Artemidorus produced five books collectively called the *Oneirocritica* that recorded hundreds of dreams and their possible meanings in the second century CE. "Of the Winds," Artemidorus wrote, "gentle winds are good, violent are wicked and evil people: troublesome tempests of wind are perils and troubles."[5] Is this how he'd interpret my storm dreams? As perils forthcoming?

Medieval Christian writers looked to the holy and the demonic as possible sources for dreams, believing the demon could appear at any moment, in any guise. But some clergy also advised against commoners engaging in dream interpretation; the laity were simply "spiritually inferior," as scholar Isabel Moreira writes.[6] Dreams were also important in Islam, with famed Muslim scholar Ibn Khaldūn categorizing them into three broad types: those from Allah (clear in their meaning), those from Angels (requiring an understanding of signs and symbols), and those from Satan (confusing dreams).[7]

More recently, Sigmund Freud—who was familiar with Artemidorus—cemented his reputation and the tenets of psychoanalysis with the 1900 book *The Interpretation of Dreams*.[8] Psychoanalyst Carl Jung showed a fascination with what dreams could tell us about our personality. Today it's possible to buy stacks of dream interpretation dictionaries, promising to provide their readers with personal insight or better sleep or access to past lives.

According to more recent neurology research, my storm dreams could be evidence of a phobia, or remainders of traumatic experiences.[9] They might also be a way of allowing me to practice for the eventuality of finding myself in one of these frightening scenarios. That theory is known as "threat simulation," and hypothesizes that nightmares are like the video games of our subconscious, helping us navigate challenges free of the risk of injury.[10]

Any of those explanations might contain an element of truth about my storm dreams. I *am* afraid of storms. I have gone through dangerous experiences in them. But I know there's a deeper layer to these nightmares. They reflect a more elemental fear: the possibility of unpredictable, unpreventable bodily harm. They're a reminder that the world can be unmade by death in the matter of a moment.

My earliest memory of an actual storm comes from the summer of 1994. My parents had decided on a final summer sailing trip up to Pelee Island, the largest island in Lake Erie, just across the Canadian border. It was the weekend before I was meant to start kindergarten.

I loved the island for its wildness. A small number of people lived there, some of them working at a winery my parents liked to visit. Normally we anchored our boat off the southern tip, Fish Point, but on this visit we headed to Scudder Marina, where we were plagued by gray skies. By mid-morning on Sunday, waves pitched themselves over the shoreline in an unceasing barrage. My parents determined that between the high seas and heavy wind, the weather was too rough for us to sail back to northern Ohio. They called the school to explain why I'd be missing my first day.

Thunder, lightning, and wind were the first metrics by which I learned to measure the danger of the sky, because these create the language sailors must learn to read. A healthy breeze made for perfect sailing, but the line between brisk and brutal could be rapidly breached. On occasion, a storm would bubble up when a race was already underway, at which point the adults monitoring us in powerboats would come by with loudspeakers to announce the event was ending early. We sometimes took down the masts of our little vessels to minimize the risk of becoming lightning rods, then joined a chain of sailors attached by the bowline, all being tugged back to the harbor by one of the safety boats.

I spent so much time on the lakeshore that I inevitably witnessed the extremes of storms. A waterspout lifting out of the water and connecting to clouds, which I gaped at from the safety of our sailing club. Sitting on the deck of our boat, witnessing another sailboat's mast get struck by lightning. The blinding flash was so quick I might've doubted it if not for the puff of black smoke and sparks that appeared in the aftermath. Watching the sky go dark as a bruise, and the water below it turn seafoam green, which augured a serious storm. Memorizing the phrase, "Red sky at night, sailor's delight. Red sky in the morning, sailors take warning."

Over the years, I learned to trust myself and my abilities, to feel mostly in control of my little sailboat even when conditions were challenging. When one of my closest friends became my crewmate, we reveled in wind and waves, her attached to a harness and trapeze, bracing her feet on the edge of the boat and leaning out over the water while I steered. Her weight counterbalanced the force of the wind on our sails, and she would grin at the movement of our boat as we caught up to other sailors on the racecourse. I could even laugh at the experience of shooting down into the trough of waves, dipping so low that all the other sailboats around us disappeared. Knowing how to stay one step ahead of the weather made me feel brave, even powerful. It meant I could handle the atmosphere's unpredictability. Accurate predictions and adequate preparations took the force out of fear. Together, I believed they were enough to keep me safe.

Thunderstorms on Earth require several physical components (such as lifting and atmospheric instability), but the most important ingredient is moisture. Our planet has had water for more than 4 billion years, and oceans formed 3.8 billion years ago.[11] With water on the earth's surface, combined with greenhouse gases creating an atmosphere that trapped heat, the potential for storms arose. There was much more carbon dioxide in the atmosphere at that point in time, meaning higher surface temperatures.[12] Another component: lightning. Today the planet might see around 560 million lightning bolts per year, whereas models suggest early Earth had 1 to 5 billion flashes annually. Scientists who have modeled the effect of these frequent lightning storms have even suggested that the strikes unlocked more phosphorous, an element necessary to the formation of life.[13]

At any given moment, there are about two thousand thunderstorms happening all over the surface of the planet; the United States alone has about a hundred thousand of them every year. But predicting when and where these storms will happen, as well as how severe they might become, is a game of chance, even with hundreds of satellites scanning the Earth's atmosphere at different altitudes all around the globe. And the effects of climate change seem poised to make these predictions even harder.[14]

Consider a location that has some of the most frequent storms anywhere, the "electrical storm capital of the world"—Lake Maracaibo in Venezuela.[15] The lake receives water from the Catatumbo River, opens onto the Caribbean Sea to the north, and is hemmed in by mountains on three sides. The end result: 250 bolts of lightning per square kilometer per year, and lightning storms occurring almost three hundred days a year.[16] According to NASA, only ten minutes of Catatumbo lightning could power all the lightbulbs of South America.[17] The incredible frequency of the storms has allowed researchers to create models that forecast the storms up to *three months* in advance. But this is truly an outlier: for most of the world, highly accurate forecasts only go out to four days.

The atmosphere is filled with such a complex array of variables that weather is known as a chaotic system: a slight change in the initial conditions can lead to a vastly different outcome.[18] Colloquially we sometimes call this "the butterfly effect," discovered through the research of MIT meteorologist Edward Lorenz. Even before the idea was scientifically proposed, there was plenty of sci-fi speculation about the ways one small change affects major events. Ray Bradbury's short story "A Sound of Thunder" (1952) employs a butterfly as a narrative device: when a time-traveling character accidentally crushes a butterfly in the distant past, it produces significant changes in the present.

It's not hard to understand our fascination with the concept: it turns our human actions and choices into a series of mysterious crossroads. What the popular imagination sometimes gets wrong is that chaotic systems do follow physical laws and are predictable—they're just very challenging (if not impossible) to predict because of the number of variables involved. As Lorenz writes regarding the variables involved in weather, "The number is truly enormous."[19] A thunderstorm on the plains of Iowa might produce nothing more than heavy rain, or it might result in a tornado. Knowing the difference in advance is extremely difficult. While tornado sirens undoubtedly save lives, the false-alarm rate is about 75 percent.[20]

For the first twenty years of my life, tornado sirens were a regular staple of the spring and summer. Ohio has just twenty-three tornadoes per year on average, far fewer than Texas, Oklahoma, Kansas, and other Midwestern states. But in November 2002, one of those tornadoes hit my small town of Port Clinton, splintering dozens of homes into fragments, knocking several into Lake Erie, causing millions of dollars in damage, and resulting in a full week off school. Miraculously, despite the destruction, only a few people were injured; not a single person died.

My family lived on the peninsula of Catawba Island and my grandparents were over for dinner. The unseasonable warmth meant my dad was outside grilling when the sirens went off. Nerves began prickling my skin almost at once. We had a large basement; wouldn't it be better if everyone waited the storm out down there, just to be safe? But my dad insisted he'd finish cooking dinner, saying he'd come tell us if he spotted a tornado, never mind the fact that nightfall was already well advanced. I was forced to sit on my hands, trying to hide my fear, hoping that this would be just another storm in the already long list of thunderstorms I'd experienced.

We were lucky. Our house remained outside the direct path of the tornado, so we didn't have any damage to speak of. Thirteen tornadoes were spawned by the storm front that spread across northern Ohio that night. The tornado that hit Port Clinton was one of several F2-strength tornadoes, meaning windspeeds ranged from 113 to 157 miles per hour, strong enough to demolish mobile homes and tear down trees. A town near us experienced the sole fatality, caused by an F3 tornado, whose winds speeds can reach 206 miles per hour.

The day after, as we drove into town, I gaped at splintered houses and mangled trees. Wind strong enough to pull an oak up from its roots, to bulldoze a timber-framed house and its concrete garage, dropping fragments into the lakeshore. How did my dad avoid injury standing outside over the grill, in our neighborhood that was named for the abundance of tall pines? If the tornado had dragged itself over us, would the house have stayed upright? Would there have been time to dash to the basement before the destruction commenced?

How are you supposed to go on living a normal life when at any moment you might die?

I wasn't sure if any of my friends spent summer nights listening for sirens, wondering what it might feel like to die being crushed by the shrapnel of a collapsing house, or skewered by a lightning bolt hotter than the surface of the sun. If I mentioned these worries to my parents, they told me not to think about things like that. We were safe; the weather was nothing to fear.

I didn't know if they were lying or if my own mind was to blame. I was pretty sure there was good reason to be afraid, but I didn't want to be a coward, a worrywart. I tried to corral my fear into one corner of my mind and to do as they said. To not think about all the ways my small, fragile body might someday come up against the unfathomable force of a wild storm.

Of course, the day my heart went wrong was perfectly sunny.

It happened at the tail end of my first year in college, as my boyfriend and I walked back to my dorm following a late breakfast. Exams were nearly over; I was getting ready to return home for the summer, where I'd intern at the local paper and coach the junior sailing team. I'd already started feeling some of the symptoms I'd later discover were a result of my disintegrating thyroid, but it would be another three years before I had a diagnosis of Hashimoto's.

I can't remember what my boyfriend and I were talking about, or how I planned to spend the rest of the day. All that's clear now is the moment I stumbled on the concrete sidewalk. That motion of taking a shortened step forward to avoid a fall seemed to jar my heart out of rhythm.

The sudden hammering was deeply uncomfortable, and as it continued, the discomfort turned to fear. My heart was a freight train careening off its tracks, knocking away at my chest as if trying to escape. Forcing air into my lungs became an unnatural activity, requiring the constant reminder to *breathe, breathe, breathe.* I made it into the dormitory then fell over on the staircase, gasping, lightheaded, not sure how to communicate what was happening.

"Something is wrong with my heart," I told my boyfriend. "I need to go to the hospital."

He drove me there as I wheezed around my heart's wild flapping. I breathlessly explained what was wrong at the front desk and was immediately taken back, despite other patients waiting with their own injuries and illnesses. Once I was undressed and reclining in a hospital bed, doctors hooked me up to an IV and covered my chest in the sticky pads of EKG wires. A quick overview of my symptoms resulted in the doctors hurrying to get a shot of adenosine. They didn't explain what it did before administering the medication. They didn't say that it would momentarily stop my heart. All I felt was a burst of heat through my arm, then pressure slamming into my whole body as intense as if I'd been knocked flat by a truck. If I'd looked at the heart monitor, I would've seen it flatline, but my attention was entirely preoccupied by the awfulness of the bodily experience.

Then—the return of a normal rhythm. In less than ten seconds, I'd been put right again. The storm in my heart was over.

Heady relief came in the aftermath. I hadn't died. My heart had returned to its normal pace, even if my pulse remained a little fast from the experience. But now that it no longer felt as if my heart might break free of my chest, a swarm of questions arose. What had happened? Was that the precursor to a heart attack? Some kind of cardiac seizure? It felt like nothing I'd ever experienced in my nineteen years of life, even though I'd been an athlete for ten of those years and had regularly pushed my body to the limits of its aerobic endurance. I knew how it felt when my heart was pumping heavily to spread oxygen to my muscles. This had not been like sprinting in the pool, or chasing down a volleyball, or lifting heavy weights.

Although it was clear the emergency was over, the ER doctors said I'd need to see a specialist to figure out what had happened. They said I'd experienced an arrhythmia, but they didn't know exactly what kind. Nor could they say if it would ever happen again or what might trigger it. The only concrete instruction they could offer was a referral to another doctor, and the directive that

I return to the hospital right away if something like this ever happened again.

They didn't tell me if this might kill me, but the implication was there. For years, I'd feared death from an external source: a car accident, a storm, a mass shooting. Now it seemed more likely that my own body would be the source of a sudden end; that death might emerge from within to sweep me under.

BROKEN BODIES ELECTRIC

"Love shook my heart, / Like the wind on the mountain / Troubling the oak-trees," wrote Sappho, a famed lyricist who was born sometime around 610 BCE on the island of Lesbos and is known for her love poems to other women.[1] While we only have fragments of her work today, her acclaim was such in antiquity that she was often referred to simply as "the Poetess," a feminine complement to Homer, whose sobriquet was "the Poet."[2] Like her contemporaries, Sappho saw the heart as a source of love, as well as the seat more generally of all the strongest emotions.

Ancient Egyptians were the first to record their speculations about the heart's physiological function, as well as its religious purpose (recording one's deeds for afterlife judgment). Greek philosophers believed it was the central hub for thought, motion, and the soul, as well as emotions.[3] Plato, for one, thought the heart was a "knot of veins and a source of the blood which races through the limbs"—not entirely inaccurate.[4] His student, Aristotle, also argued strongly in favor of this cardio-centric hypothesis, believing (mistakenly) that the heart's aorta transformed into nervous tissue. In fairness to Aristotle, his anatomical research was performed almost exclusively on other animals. The only human he dissected was a forty-day-old fetus.[5]

Given the centuries-long confusion over the heart's function and its appearance, perhaps it's no surprise that we represent this organ with a symbol that looks little like it. The now-iconic heart shape may have originated with Greek coins; many of them bore emblems of heart-shaped silphium fruit, used medicinally both as an aphrodisiac and a contraceptive.[6] Some scholars think the ideograph we know today comes from around the end of the Middle Ages, after Europeans cemented the symbol and initiated cultural beliefs about romantic love.[7] They also incorporated the heart into religious symbolism, where it could either represent Jesus's love for humanity, or original sin.

Even after centuries of anatomical investigations into the heart's actual shape and function, there's no question that we still believe the heart has a direct relationship to our emotions. This is reflected even in modern medicine: Takotsubo cardiomyopathy, "broken heart syndrome," is a sudden weakening of the heart that can occur after a severe emotional blow, such as the death of one's spouse. In rare cases, it can be fatal.[8]

What's less apparent in the way we conceive of the heart is the role of electricity. We talk about "sparks" between potential lovers, but rarely in the context of the energy required for the pumping of our blood. The heart muscle is a spiral, a whirlpool of ventricles and atria and a sinus node that generates electrical signals to travel specific pathways, which enable the heart to wring as if it were a wet shirt. After I experienced my arrhythmia, I was sent to see an electrophysiologist: someone who specializes in the circuitry of the heart. If the adenosine hadn't returned my heart to its normal rhythm (or if it had caused cardiac arrest, a rare side effect), a defibrillator would have delivered between 200 and 1,000 volts in hopes of reminding the muscle of its purpose.

Maybe instead of the centuries-old heart symbol we're so accustomed to, we should be representing our hearts with a lightning bolt. I'd argue it's more accurate as well as being more potent: nothing makes you appreciate the electrical nature of your essential organs like staring down shock paddles and knowing they might be called on to resuscitate you. Even in the realm of love, the metaphor is apt; for the French, love at first sight is "*un coup de foudre*," or a bolt of lightning.

While electricity is crucial for our cardiac health, our whole bodies are filled with electric current, though the route toward attaining this knowledge followed a similarly winding path to our understanding of the heart.[9] Natural philosophers deemed this strange force "animal spirits"; later researchers learned how to agitate the muscles of dead frogs and hanged criminals, terrifying audiences.[10] Their work inspired writers like Mary Shelley, who dreamed up a horror story of the creations that come from scientific overreach. What animated us beyond the soul? It took decades of investigation to discover neurons and to pinpoint the ways that sodium and potassium rushing in and out of the cells changes their electric charge.[11] We are bodies electric, millions of little jolts moving limbs, expanding lungs, powering thoughts, controlling the stomach's ability to empty after food has been digested. If a storm is nothing more than moisture, instability, and the clashing of hot and cold molecules, it's hardly a metaphor to say that our bodies are microclimates of controlled thunderstorms. The main difference is that lightning from storms is produced by the movement of electrons, and electricity in our bodies is created by charged atoms.

I hoped that learning more about this component of my biology would provide some solace while doctors tried to figure out what arrhythmia had sent me to the hospital. Whenever I had smaller palpitations, a type known as "premature atrial contractions" (PACs), I told myself that sometimes the electrical signals got ahead of themselves. Sometimes, a pulse of bodily lightning told the heart to contract just a moment too soon; it was nothing to fear. But those PACs occasionally triggered the more frightening arrhythmia that sent me to the hospital. And over the course of four years while I was an undergrad in Ohio, this arrhythmia happened at unexpected moments. When I was at a house party; when I was getting ready for a formal dance; when I was out on my family's sailboat, far from land. Every time I had another episode, I remembered what the ER doctors had said: seek immediate medical treatment.

But I never made it back to a hospital, either because the episode didn't last long enough, or because I was simply incapable of getting to one. Each time it happened I wondered again what was wrong with the thrum of my heart. Each time, a little more fear soured my relationship with the organ that was supposed to be a source of love. I was so scared of my own heart that I learned how to check my pulse in my wrist, to monitor my average beats per minute in hopes of catching a problem before it killed me. I knew my vigilance wouldn't truly prevent bad things from happening. But still I strived to read my body's unspoken language, the same way that the "deranged" young man in Vladimir Nabokov's story "Signs and Symbols" attempts to find meaning in all that's around him. I knew my deciphering was a fruitless exercise, but it felt like the only way to keep from being swallowed whole by the terror of sudden oblivion.

Throughout this period, it wasn't only my heart that seemed to be haunting me; it was also a series of storms. My university in southern Ohio had at least a few tornado sirens every year, which meant needing to know our shelter locations in all the different academic buildings and student housing. One of my professors, along with a small group of his students, had been struck by lightning while doing fieldwork in the Bahamas. They all survived, though many had injuries that required seeing a doctor.[12]

One late spring, the harbor in Leamington, Canada, where my parents had docked our sailboat on multiple occasions throughout my childhood, was hit by a tornado, resulting in docks being torn out of the water. That weather system generated a total of fifty-three tornadoes over the weekend, killing seven people.

Compared to heat waves, which are estimated to kill around 1,300 Americans every year, tornadoes are far less deadly, with only about 70 fatalities on average.[13] But tornadoes are a more visible, dramatic force; they ignite our imagination and electrify our nervous system. Their strength shears through the trappings of modernity: homes, schools, crops. The size and violence of them expose social inequities: people living in mobile homes or less well-constructed dwellings are more likely to lose their property.[14] And thanks to the geography of the United States, our country experiences the most tornadoes of any location in the world.[15] We're destined for destruction on a regular basis.

Fortunately, a tornado is never guaranteed when a storm hits. The air closest to the ground must be warm and moist, while higher up in the atmosphere it must be cool and dry. When wind is moving in different directions at different levels of the atmosphere, this starts to create spinning. If gusts of warm air rise from the ground at the same time that cooler air from the storm sinks down, this produces spinning near the ground. When spinning air in the storm and on the ground meet, they connect to form a tornado. Often, when this type of spinning air is spotted on radar, a tornado warning will be issued.

Whether tornadoes are becoming more violent or frequent as a result of rising temperatures is difficult to answer. Intuitively, one might think that tornadoes, born of moisture in the air and hot and cold fronts clashing, would have more energy at their disposal to draw upon with all the impacts of climate change. But as we've seen, the factors that lead to the formation of a tornado are slightly different from those of a simple thunderstorm. Researchers are still trying to understand all the variables at play, from vertical wind shear to convective available potential energy to the changes that occur between La Niña and El Niño years (conditions in the Pacific Ocean that affect the weather worldwide).[16] Making the task even more challenging is the lack of good records. Until the 1990s, the National Weather Service relied almost exclusively on someone seeing a tornado and reporting it, which meant the tornado was generally large and longer-lived. Today, Doppler radar can detect even weak tornadoes that disperse after a few minutes, which means that any increase in the recorded number of small tornadoes is more a reflection of our technology than proof that their frequency has changed.

While some scientists think warmer temperatures will result in more thunderstorms, how that will play out in terms of tornadoes spawned is much harder to say. There are some indications that tornado outbreaks (periods of one to three days with at least six tornadoes) are getting larger and more frequent.[17] They also seem to be moving into different locations beyond "Tornado Alley," the stretch of central states from South Dakota to Texas that have historically experienced the most tornadoes. And scientists' models suggest that the frequency of lightning will increase by an average of 12 percent for every degree of warming the planet experiences.[18] But we need more research to understand all the dynamics at play in our storm-laden skies.

Medicine faces a parallel complexity problem. A body is a vast collection of proteins and cells and organs, overlapping systems in which a microscopic malady can undermine the whole. Parsing the difference between correlation and causation is rarely straightforward. In my own case, no doctor can say with certainty how much my nascent thyroid disease may have contributed to my heart problems. An absence or abundance of certain thyroid hormones may trigger palpitations and arrhythmias.[19] But there are people who have heart palpitations without having a corresponding thyroid disease, and vice versa. Like the atmosphere, our bodies are chaotic systems, in which minutely different ingredients can result in widely disparate outcomes.[20]

If I'd been diagnosed with Hashimoto's thyroiditis at the same time as I first began having heart arrhythmias, maybe doctors would have identified the specific type of arrhythmia sooner. Maybe I wouldn't have had so many episodes. Then again, maybe not. This hypothetical is a Gordian knot, a puzzle I'll never unravel because there's too much I don't know. I wish I could have spared myself from some of the suffering, from that burden of fear. I wish I didn't have to go through the experience of becoming estranged from my body and relearning its nuances, over and over again.

Medicine and meteorology alike attempt to provide humans with more information, more advance warnings, better and longer lives. But there comes a point when even the experts have to say, "We don't know." To be alive on this planet requires making peace with perpetual uncertainty. At age twenty, I didn't understand how to do this without being permanently afraid.

For the entirety of my undergraduate years, I tried to control the uncontrollable. Breathe through panic when the heart palpitations struck. Hide in dormitory basements when tornadoes approached. Don't drive in storms, don't drink caffeine, don't think about death. But life thwarts all illusions of control. The latter half of 2012 brought my heart and storm fears to a zenith, as if all my resistance had only built up enough pressure for a series of explosions.

The first storm came on a sunny July afternoon, one day before I was slated to move to New York. It had been four months since my last arrhythmia attack; in two months, I would be diagnosed with a thyroid disease.

My partner and I were driving to his parents' house on Catawba Island after a trip to the beach. Clouds started lumping across the sky while we were still wandering Lake Erie's shoreline, but they didn't attain an ominously dark hue until we were in the car. Less than ten minutes from his childhood home, golf ball–sized hail started pounding the roof of the car. I felt the cacophony in the hollow of my chest, as if I was standing next to the speaker at a concert. But it still wasn't loud enough to drown the tornado siren that sang out moments later. The wind whipped so ferociously that the SUV rocked. My partner gripped the steering wheel and I clutched his knee, mirrors for one another's alarm.

As we sped into his family's neighborhood, enormous oak and maple trees toppled around us. A large trunk crashed onto the road, completely blocking our path. When my partner hesitated, I shrieked, "Drive around it!" He sped through a neighbor's lawn, the tires leaving wet grooves in the grass. Power lines collapsed; the windshield wipers barely managed to keep up.

When we pulled into his family's driveway, the garage door creaked open and we ran inside, both doubled over to get in as quickly as possible, dinged by the hail and the confetti of green leaves torn from their trees. I paced between the kitchen and living room, breathing hard and shaking so badly I couldn't speak, feeling the occasional small heart palpitation. The adrenaline took twenty minutes to leave my system; the storm lasted barely half as long.

Later, when we ventured back outside to survey the damage, meteorologists announced it had been a "microburst," a thunderstorm that hit one small segment of the little peninsula of Catawba. These storms can be as devastating as tornadoes, with winds up to 150 miles per hour. We wandered the streets, gawking at the tangle of collapsed trees and power lines, the piles of hail glimmering in everyone's yard. Multiple cars had been crushed. I couldn't comprehend the destruction we'd passed through unscathed. If we'd left the beach three minutes earlier, would one of those enormous trees have killed us? If we'd stayed at the beach even longer, would we have been beyond the reach of the storm?

I tried to make myself forget. I knew it was useless to worry about the weather. In any case, we were moving to New York City, where tornadoes were exceptionally rare and we wouldn't even own a car, let alone be caught driving one during a storm.

But only three months later came the hurricane. Evacuation orders trickled out of the mayor's office the day before Sandy made landfall. My apartment was only a block outside the evacuation zone, and my partner was away traveling with family; it seemed safer to stay with a classmate on the Upper East Side than remain alone. All that long night trying to sleep on my friend's couch, I listened to the wind banging around the fire escape. With morning came news of the calamity: power off for dozens of blocks; flooding all around the five boroughs; hundreds of houses destroyed; subway lines submerged; forty-four people dead.[21]

Two weeks later, as some New Yorkers were returning to a semblance of normalcy while others still slept in shelters, my heart went wrong again. Was it related to the storm? Research has consistently shown that heart attacks surge after people live through a natural disaster, but my arrhythmia is not the result of cardiovascular disease and has little in common with myocardial infarction.[22]

I decided to see a doctor. My partner and I walked ten blocks to a Manhattan urgent care, my pulse racing at 150 beats per minute. The booms at my collar bone and throat made breathing feel as if my lungs were a rusted bellows.

In the undecorated patient rooms of the urgent care, the after-hours doctor called for paramedics. By the time the team arrived, my limbs were seizing up, arms locked in a bird-wing position. The EMTs pulled down my forearm to expose the veins near my elbow. But spasms still electrified my muscles. The first attempt to insert an IV resulted in a spray of blood shooting from my arm, covering my shirt, my bra, the floor.

"What's happening to me?" I tried to ask. But my face had gone numb. The words were an indecipherable slur. This inability to speak, to make myself understood, scared me more than anything else. I squeezed my eyes shut and tried to picture my partner, my family, my closest friends. If I was going to die, I wanted to be thinking of them.

Eventually the paramedics managed to insert a needle into my vein. But the first dose of adenosine failed to break the arrhythmia. At that point, they administered a second one. This time, the momentary cessation of a heartbeat succeeded in reestablishing the proper rhythm. The tympanic drumming ceased. My atrium slid back into its familiar rhythm. I pried my eyes open to see that I was being taken out to an ambulance.

There followed a sleepless night with blood tests and heart tests and potassium pills to bring my blood chemistry back in balance. "Supraventricular tachycardia," the doctor declared. I'd need surgery to correct it. For more than three years I'd been struck by these episodes without knowing what to call them. To have an answer, a path forward, a possible cure felt like learning an incantation, a string of words I could call upon to protect myself from the unknown.

I was discharged into a gray sunrise, the streets cold and damp. Fifteen blocks south of my apartment, the electricity had only come back on a week ago. Flooded subway lines would remain unusable for months. Some of my classmates hadn't been able to get to the northern end of Manhattan and attended class remotely. Surely my professors wouldn't mind if I missed a day; we were all in various states of emergency. I sent off several emails before getting into bed, closing my blinds against the growing light.

I made it through the rest of the fall semester without talking about my heart problems. I was too scared by the enormity of it, by the thought of dying suddenly at age twenty-three. It was easier to focus on my reporting assignments, nearly all of which revolved around the hurricane. I repeatedly visited one of the Brooklyn bases of a volunteer movement, Occupy Sandy, where tens of thousands of people answered phone calls and accepted donations and drove around the city to make deliveries.[23] I conducted interviews, took photos, helped organize boxes of diapers and powdered milk and canned beans. Some of the volunteers had participated in the 2011 Occupy Wall Street movement; they talked about this new iteration of Occupy as an anarcho-syndicate, a group with a horizontal power structure that would encourage citizens to recognize systemic injustices in our government and fight against them.[24] Others had volunteered in New Orleans after Hurricane Katrina and wanted to contribute their knowledge and skills.[25] Still others came to help simply because they saw how much help was needed.

One day I rode with volunteers out to Midland Beach on Staten Island, one of the worst-hit neighborhoods. Even three weeks after the stormwaters receded, streets were still covered in sand and a truck was rolled on its side. Electricity for the neighborhood came from diesel-powered generators; no streetlights broke the early fall of darkness. Families who had requested assistance cleaning out their flooded houses were wary when we first approached, then voluble when they realized who we were. They told stories of being rescued by police and losing most of their possessions and listening to the crash and thud of waves against their doors. A man whose wife had the terminal disease ALS was on a ventilator; her machine had batteries that were good for eight hours, and the couple's power was out for three days.[26] The husband looked one place after another for generators to keep her breathing.

What if my incident of supraventricular tachycardia (SVT) had happened the night of the storm instead of ten days later? Would I have called an ambulance? Tried to walk to a hospital in the fierce wind? A young couple not far from my friend's apartment was killed by a falling tree branch the night of the storm while out walking their dog.[27] That could've been me. It could've been any one of my friends or professors. Another devastating storm could—probably would—hit New York City again. Another bout of SVT might engulf my heart at any moment.

Had I lived thousands of years earlier, before medical intervention, before meteorological explanations for hurricanes, what would I have drawn on the damp walls of a dark cave, with nothing but fire to create the alignment between hand and eye? Maybe a black sky, a tremendous wind, a silhouette of a body riven in two by an errant bolt of self-generated lightning.

I scheduled my surgery for winter break but told hardly anyone that it was happening. My heart was almost unbearably calm. It reminded me of a cloudless day shortly after the hurricane that felt more like a taunt than a moment of respite. My body was too unsettled to accept repose.

The nurse prepping me shaved around my groin and struggled to place an IV in the back of my hand, the vein releasing a torrent of blood when the needle penetrated. She commented on my skinniness, asked if I was a model, finally moved to a vein in my elbow. The last bit of prep was the sticky defibrillator pads pressing against my back and chest, reminders of the worst-case scenario. Death was unlikely, but never impossible. Slightly more likely was that my heart would be damaged during the operation and I'd need to have a pacemaker installed. For weeks I obsessed over those eventualities, scaring myself even more.

I had to be awake for the procedure, because general anesthesia can suppress the very arrhythmia the surgeon was trying to track down. I was hyperventilating by the time they wheeled me back, despite the anti-anxiety meds dripping through my IV. The surgeon cut two blood vessels near my groin, then inserted sheaths and wires that wound through my abdomen. Once they reached my chest, the wires delivered shocks to speed up my heartbeat and determine the source of my arrhythmia. It felt as though someone were gripping my heart, forcing it to beat. My heart rate climbed to three hundred beats per minute—so fast it was no longer distinguishable as a pulse, becoming an unbroken vibration. When the sensation ended, a new one commenced: the SVT, now strangely familiar and comfortable after what they'd been doing to my heart.

A new burning sensation followed shortly after. "What is that?" I asked.

"Don't talk," the doctor said. "We're burning the pathway." A scar would form across my heart and impede the faulty electrical signal from being triggered. If successful, this surgery would mean my heart could no longer beat its inappropriate rhythm once the wound healed.

In the following days, my heart jumped into its arrhythmia for short periods, then subsided, a pattern the doctor had warned me about before I was discharged. Walking was painful due to the incisions, but they healed quickly. I was tired, and then less tired, and then the SVT stopped happening. I thought—*I'm actually cured*. For the first time in years, I could trust the steady beating of my heart.

I thought maybe fixing my heart would mean fixing *all* my fears of sudden death: no more worrying about tornadoes or hurricanes or microbursts or being in the wrong place at the wrong time. Any of those things might still happen, but my heart had been restored, so now I wouldn't be afraid of them. Along with my new heart would come an infusion of courage.

The trouble with fear is that it's meant to protect us. When our animal bodies are convinced of danger, cortisol and adrenaline help us stay alive. And I'd lived through all those storms and heart problems, hadn't I? The fear had done its job. Even if I wanted my mind to embrace a new reality, one that laughed at the idea of fear, my body had been taught otherwise. Brains can be warped by severe or prolonged exposure to fear, as we see in PTSD. The body records the disasters it lives through, sometimes by altering genes. The DNA itself doesn't change, but the expression of those genes can. This is the field of epigenetics, and research has shown that fears and traumas are passed down, inherited by a new generation who has yet to face the horrors of mortality, of all that comes with life.

If we break stories into symbols, like Joseph Campbell, we come to the hero's journey, that call to leave what is known and comfortable to face trials and tribulations. At the end of the adventure there's a moment of homecoming. But sometimes the hero is too transformed by their travails. Frodo Baggins doesn't stay in the Shire after bearing the evil of the One Ring and facing the might of Sauron; the hobbit is no longer at ease in his home.

My most fundamental home has always been, and will always be, my body. But I wasn't sure if it would ever feel welcoming again, not after my thyroid's self-destruction and my heart's erratic behavior. There would be no return to safety, not while I still carried so much fear of the future, not while I mistrusted my own flesh.

BEFRIENDING FEAR

Three brothers were born from the gods Gaia and Uranus, a trio collectively known as the Hecatoncheires. Each was as massive as a mountain with fifty heads and one hundred hands. Together with their siblings, the Cyclopes and the Titans, the first children of Gaia were imprisoned below the earth because Uranus, their father, feared them. But Cronus the Titan broke free with his mother's help and overthrew Uranus. He released all his siblings except the Hecatoncheires, who remained entombed in darkness. It wasn't until Zeus came along to lead yet another revolt, this time against the Titans, that the hundred-handed brothers breathed fresh air once more. By siding with Zeus and the Olympians, the Hecatoncheires secured their liberation and a permanent place in the mortal world. The makers of destructive hurricanes, they would be unleashed at the gods' whims whenever humans displeased them.

So goes one telling of the story. But there's another version, a lost Greek epic called the *Titanomachy*. Little is known about the text, since only a few fragments survive. It has been attributed to a person named Eumelos.[1] The subject, however, is clear enough from the title, which means "War of the Titans."

I've always loved the idea of lost texts, not only because they offer the tantalizing possibility of rediscovery, but also because their absence produces fissures in what's known. As long as a text remains lost, it could be said to contain anything. In my imagining of the *Titanomachy*, the Hecatoncheires are sisters, deemed too monstrous and powerful for freedom. After all, the Greek poet Hesiod described the Hecatoncheires as "too terrible to be approached or even spoken of," writes classicist Debbie Felton.[2] The Cyclopes were another trio born of Gaia and Uranus, deadly in their own ways, but the one-eyed brothers seemed not to have inspired the same level of fear. A monstrous man can be a weapon, a tool. A monstrous woman refuses to be put to any purpose but her own.

Maybe in this lost version of the tale, Zeus has the same fears as the Titans: that the hundred-armed mountains could unmake the world. Perhaps he left them locked up, and the sisters engineered their own escape. Maybe at the end of the *Titanomachy*, it's not the Olympians who ascend to power; it's the Hecatoncheires. I imagine them wearing dark shrouds, hair flying around them like scraps of cloud, their arms cradling water and wind, their voices the grumble of far-off thunder or booming clashes, depending on their moods. They roam the planet, never silent, never again trapped.

My version does not fit with what we know of the Hecatoncheires from the *Titanomachy*. But it's unlikely that we'll ever find the full original version of the *Titanomachy*, written nearly two thousand years ago.[3] Some legends are simply lost to time.

Science though—it has produced tools and methods that let humans pry deep into the past. The field of geology sees researchers reading Earth's layers like a book; some of those scientists have found evidence of storms dating back millennia. Researchers in the field of paleotempestology exhume sediment in coastal marshes, which contains layers of sand and microfossils among the finer muck—evidence of the fury of ancient hurricanes.[4] Stalagmites in caves carry markings from severe flooding, the kind that can only be caused by massive storms.[5] Even more astonishing are fulgurites, scars of glass formed by bolts of lightning. They hold chemical clues about the environment in which they were formed tens of thousands of years earlier.[6]

When I think of the weather, I envision this great lineage. A meeting of earth (Gaia) and sky (Uranus) that happens over and over again, displacing the first progeny of their union to make room for new rulers. No species' reign is ever permanent: Cronus overthrew Uranus and Zeus overthrew Cronus. Over the course of its long history, our planet has sheltered predatory ocean worms, tree-sized ferns, insects with wingspans greater than two feet, synapsids, dinosaurs, mammals. And one of the few constants across the eons—no matter who or what may believe themselves to be the "chosen," the "destined," the "rulers"—is the ability of weather to wipe away whatever version of the world currently exists.

Sometimes we fear monsters because they're unknown; sometimes because their incredible abilities are obvious, but the use to which they'll put them is not. A monster might decide we're nothing more than meat. A monster might have no thoughts about us at all, simply moving where it chooses with little regard for the destruction it creates. I know logically that a storm has no particular animus toward humans. It's a product of natural forces. But I also understand why so many human cultures have made storms into gods or monsters and looked for ways to placate them. I understand the impulse to pray, even when you're not sure what good it will do.

At first, it seemed that my heart surgery had worked. For years, I had no episodes of supraventricular tachycardia. I still had the light, fluttering palpitations called "premature atrial contractions," but they happened infrequently enough that I could ignore them. They occurred most often in the week before my period, or when I was exhausted, or falling ill, or experiencing intense emotions. I told myself that my heart was easily irritated, that having a thyroid disease probably didn't help, that the real problem had been done away with. When an endocrinologist ordered me to do an echocardiogram one year after my surgery, he told me that there was a problem with one of my heart's valves, but that it was a benign deformity. None of my cardiologists had ever mentioned this, despite my having done the same exams with them, so I decided it wasn't something worth worrying about.

Until the palpitations got worse again—much, much worse.

The first time they kept me up all night was after a dinner of Spanish rice and beans. The second time was at the tail end of leaving a job and moving. The third time was following knee surgery for a torn meniscus, my first experience with general anesthesia. This third time, the palpitations didn't stop. Night after night they came, keeping me awake till, delirious with exhaustion, I fell into a doze and dreamt I was some enormous spaceship being shaken to pieces in an interstellar battle. I woke up and went to the hospital.

I thought I'd exorcised my arrythmia. It was true that any time these premature atrial contractions triggered a run of SVT, the old arrhythmia ceased after a few breaths. The scar tissue on my heart was still doing its job. But a specter in the upper chambers, the atria, was now restless. Instead of allowing the sinus node to regulate the rhythm of beats, atrial spasms inserted additional beats, resulting in a sensation that varied from a butterfly-wing flutter to aggressive booms that speckled my vision with dark stars.

Virtually all people experience these ectopic beats at some point or other, but not everyone feels them, and most people don't have dozens of them in the span of a minute. The frequency with which mine happened—hundreds of times an hour, an unpredictable cannonade erupting behind my ribcage—was worrisome. But ER doctors could offer no remedy once they confirmed I was not actively experiencing a heart attack. I began to learn what so many chronic illness patients know: the hospital is largely unequipped to handle the medical needs of complex patients, the kinds of problems that aren't quite life-and-death but are also far beyond the norm.

Back to an electrophysiologist once more. This time the option of cardiac ablation wasn't even offered. Locating the origin of a premature atrial contraction, or PAC, is like playing a rigged game of whack-a-mole; there is too much atrial real estate available to the pushy beats. May as well burn away the whole heart in search of errant nodes. But the doctor felt something had to be done, as the PACs were severely disrupting my life. It was unlikely they would kill me, but having them at a high frequency increased the possibility of developing other arrhythmias, like atrial fibrillation, which can lead to strokes.

The doctor prescribed beta blockers, a medication that lowered my blood pressure so I wouldn't feel the PACs quite so intensely. The downside was that I already had quite low blood pressure, so the pills sapped all my energy and made me liable to pass out.

In the midst of my struggle for a quiet heart, my grandmother was dying. When I was in elementary school, Grandma B. lived with my family for a while after her husband died. She taught me a bit of French (her first language), and how to play gin rummy and pinochle, and how to make flaky pie crust for her incredible apple pie. Dementia stole most of those memories from her before the end, but she still knew who I was when I visited her in hospice, when I wet her dry lips with a pink sponge on a stick. Her fingertips were deep purple. She held my hand and told me she'd pray for me; a lifelong devout Catholic, she felt prayer was her most powerful form of care. She'd told me once that she was scared of dying, but now she only seemed tired.

Grandma B. never knew of my heart troubles, not the SVT or the PACs. My parents instructed me not to tell any of my grandparents about my heart surgery because they'd only worry. I knew it was especially true of Grandma B.; it seemed a mercy to spare her some of that concern. But as I sat with her while she labored to breathe, a never-ending chain of Catholic masses playing on the TV, I wished I'd shared my own fears of death. I wished I'd asked her how she maintained the courage to go on despite dealing with epilepsy, and multiple autoimmune diseases, and, later in life, a series of small strokes. *How do you handle being so scared?* I wanted to ask. *When does the fear let you go?*

I went back to my hotel room that night and cried to my partner, because I wanted to devote all my mental energy and emotions to my grandma, but the ongoing heart palpitations stole some of my attention. I didn't know if the cowl-hooded figure coming for her might decide to take me, too.

Maybe her prayers worked. She died peacefully—if I can say that without having been there to see it—and within another month my PACs were finally reduced enough that they became no more than an occasional menace. I'd survived this round; but so, too, had my fear.

Although scientists haven't yet deciphered how increasing levels of greenhouse gases will affect tornadic storms, they're clearer on the future of hurricanes. It's not a future anyone is excited about. Melting ice caps will lead to higher sea levels; higher temperatures in the air will create higher temperatures in the water; increases in both water and heat provide greater energetic power for hurricanes to draw upon. Even if the storms don't become more frequent, they are likely to grow more intense, a trend that is already happening.[7] Atmospheric scientists have suggested we need to expand the scale on which we measure the strength of hurricanes, adding a "category 6" for the most potent storms ever recorded.[8]

Unsurprisingly, the thought of facing such storms, along with the other ravages of a climate-disrupted planet, has made people nervous about the future. The concept of "eco-anxiety" has begun to permeate our culture, especially among young people.[9] Some psychologists have described this as a form of pre-traumatic stress disorder and suggest that such feelings are a normal response to the ongoing crisis. Researcher Britt Wray notes in her book *Generation Dread* that fear, anxiety, hopelessness, and horror are all appropriate responses to the many-layered catastrophes we witness around the world. She writes, "It is *sensible* to get spooked when a group of leading environmental researchers publish a paper that opens with the words, 'The scale of the threats to the biosphere and all its lifeforms—including humanity—is in fact so great that it is difficult to grasp for even well-informed experts.'"[10]

The fear of a disastrous, dystopian future is a reasonable one, but that doesn't mean we should abandon any efforts to change things. Despair is a luxury none of us can afford. Instead, Wray suggests, fear is something we should face and understand, then move through. We may always carry painful shards of anxiety within us, but it doesn't have to cause paralysis. Even at our most frightened, we can still exert some agency.

When my boisterous PACs returned yet again, several years after my knee surgery, I succumbed to panic attacks. I went, reluctantly, to the hospital, because my chest was beginning to hurt, and breathing had become more difficult to do around the many short runs of SVT and sporadic ectopic beats. Once again, no signs of a heart attack or imminent death. "Follow up with an electrophysiologist," the ER doctor instructed.

In the days between that hospital visit and my appointment with a new cardiologist, the PACs continued unabated. They were distracting, a constant rumble of uneven thunder that might calm for a few hours before erupting once more. At some moments, it seemed my anxiety had worn itself out, like a child's tantrum ending with a nap. When I could sit with the erratic beats and loosen my other muscles of any resistance, I could pretend it was like watching a thunderstorm from the front porch, the scent of petrichor rising from dampened earth, a hint of spray fizzing along my arms. Surrounding me, but not consuming me.

Even better than the brief moments of tranquility was my new electrophysiologist. He spent more than an hour reviewing my full medical history, then hooked me up to an EKG machine until he could capture multiple ectopic beats, which no doctor had ever done before.

"I can imagine you sitting at the proverbial typewriter and getting these, then not being able to concentrate," he said. He ordered an echocardiogram to check on what a previous endocrinologist had thought was a valve defect, only to discover there were no structural problems. He also prescribed an anti-arrhythmic medication for me to use for several months to try to reduce the PACs. If the medication didn't work, ablation was an option, though the rates of success were highly variable. That said, if I was having the beats frequently enough, it might be easier to locate their source in surgery.

"Of course you're anxious and distracted," he said. "That's natural. It's like that torture of having water drip in your face for hours on end. You don't know when the next drop is coming. You can never relax."

Of course I was anxious. Not a single doctor had ever offered such empathy; in turn, I had rarely offered it to myself. The narrative I told myself of my fears was that they were a weakness, a problem. I'd never given myself grace or kindness, even though I was in desperate need of both. My heart was hardly the only errant body system occupying my thoughts.

The palpitations regularly caused insomnia, but sleep was a struggle even without them to blame, especially in the ten days before my period. I'd lie in bed exhausted but unable to lull myself to sleep, no matter how I tried. Once my cramps started, the insomnia was even worse. I'd never considered my periods to be very bad, especially compared with stories I heard from other women about bleeding through pads, vomiting from the pain, unable to leave bed for days. But over time, my periods were getting worse. The pain started feeling jagged, knife-like, and it occasionally hit at random moments in my cycle outside of when I was bleeding. At times it was so intense I couldn't speak, could only curl in around my abdomen and remind myself to breathe. I couldn't watch TV or listen to a podcast or do anything to distract myself from the pain; I needed to focus every ounce of energy on not screaming. A gynecologist finally prescribed muscle relaxants to help me get through the monthly horribleness.

All this happened alongside my heart struggles, my thyroid disease. When I had the energy, I'd research the overlap between the different symptoms my body experienced, scaring myself with the thought that there might be more maladies developing within me, like storm clouds gathering at sea that will coalesce into a violent hurricane.

And there were still storms, still that fear in me, which was especially potent whenever I happened to be in the car and thunderclouds appeared on the horizon. When tornado sirens went off while I was driving on Chicago city streets, my chest squeezed with anxiety. The awful image of being crushed or swept away by the wind played before me on repeat. When hail pounded the roof and windshield, my limbs shook and my mouth went tacky. When my partner and I planned a road trip to visit National Parks out west, we discussed our strategies for what to do if a storm occurred on the road.

Sure enough, we faced several storms on that trip. After the first, a thunderstorm that produced nothing more than heavy rain, it occurred to me that here was something else I had good reason to fear. *Of course* I hated storms in cars; a tree had nearly crushed me once. *Of course* I panicked with the sound of tornado sirens. Maybe my fear wasn't a weakness that I needed to bully myself out of or a failing that I needed to overcome. Maybe I didn't need to feel so ashamed. Nearly all humans feel fear, and none of us *choose* what will initiate that physiological response. Life conditions us. Even our genes may condition us.[11] Why berate myself for feelings I couldn't control? Why deny I was having them in the first place? I would feel the fear, then feel whatever came on the other side.

A week into our road trip, we decided to skip spending another night in North Dakota and drive straight through Montana. It would mean ten hours in the car but would put us that much closer to our destination, Glacier National Park. And because my terrible cramps were ramping up, my partner agreed to drive most of the way so that I could take a muscle relaxant.

We slid through hills on the eastern edge of the state, highway turning to two-lane roads. The land stretched on and on beneath an enormous sky. Why did that blue space above us look so much vaster than it did in the Midwest, where we lived? We zipped past cows and sheep and mule deer, occasionally talking, until I felt too woozy from the muscle relaxants to do much more than help with navigation.

It was around that time that the storm clouds began appearing ahead of us. Monstrous black things, stretching all the way down to the horizon in impenetrable sheets. Our phones alerted us of severe thunderstorm warnings for the counties we'd be entering. The forecast escalated from dime-sized hail, to quarter-sized hail, to golf ball–sized hail. Spectacular streaks of lightning split the dark sky, coming so frequently all around us it felt as if we were driving toward a holiday firework display. On these empty roads, there was nowhere for us to take shelter if heavy hail hit.

I waited for my heart rate to increase, for my palms to sweat, my hands to shake. But none of that happened. The muscle relaxants likely had some hand in keeping me from feeling perturbed, but the storms were also beautiful. They turned the sky shades of purple and blue I'd rarely seen. Watching them inspired more awe than fear. There was something ineffable in seeing the sky work magic. I tracked the black clouds and waited for fear to strike, for the darkness to enclose our car.

But they never did. Every time we came close to the storms, they bubbled around us, as if clearing a path. Each turn we made seemed, at first, to be bringing us directly into the onslaught. Instead, the clouds would roll to the north or south, breaking like waves. As the sun set and the whole sky went dark, we listened to booms of thunder and the sprinkle of rain. We drove past ranches and tiny gas stations before finally making it to the town that was our destination.

Exhausted and stiff, we checked into our hotel, where two receptionists were discussing the weather. "Hell of a storm," one was saying. "Did you see the size of the hail?" My partner and I looked at each other, somewhat agog at our luck. We'd driven into the heart of a line of thunderstorms, but somehow, we'd missed it all.

PART THREE

FLOODS—UTERUS—ANGER

WITCH HUNTS

Water has been pulling me to it as far back as I can remember. I was one of those kids always falling into creeks and lakes and marshes. On a kindergarten trip to nearby wetlands, I tumbled off the boardwalk into the swamp, sharp grasses and slimy algae infiltrating the inside of my pants. It happened at least twice in the pond behind our house: once while catching a tadpole for a friend, once because I climbed too far out on a fallen tree. I broke through thin ice while skating and sunk to my knees; I flipped over backward off a sailboat when I hiked my body out over the edge and missed the strap that I was supposed to hook my feet under; I slipped off a patch of dry grass in a flooded field filled with rainwater and cow shit. Even at its foulest, I've never resented the water for its gravitational pull on my body. My cells sing to this most essential element. Water is graceful, fickle, changeable, beautiful. I can't imagine living far from it.

Maybe that's why I so welcomed my period: not only because it came later than many of my friends' periods, but because in my fourteen-year-old imagination, monthly bleeding proved me to be another creator of water that came with the possibility of life. Not that I had any interest in pregnancy—it was the mythic symbolism of it all. Womanhood felt tied to the lakes, rivers, and oceans, the veins of water wrapped around our world.

But my vision of this melding between femininity and watery bodies is not one celebrated by most of the world. Women and water alike are powerful life-makers, but all too often that power creates fear, envy, and disgust. We become controlled and exploited. It was around the time my period started that I noticed how often my body and behavior were being hemmed in by the people around me. Being a teenage girl meant my flesh was up for grabs.

Boys snapped my bra, a painful and humiliating sting that I was conditioned to laugh at. They told jokes about how vaginas smelled fishy, which baffled me, since I'd grown up fishing. If a girl got angry or cried, it had to be because she was "on the rag" or "riding the crimson wave." I began to see that womanhood meant a continual infringement on my autonomy and emotional well-being. Womanhood was an infinite series of trapdoors through which I would inevitably fall, no matter how carefully I stepped.

There was the boy who frequently followed me around the YMCA before swim practice, stealing my schoolbooks when I tried to do homework, which forced me to hide in the women's locker room with its uncomfortable wooden benches unsuited to writing. Another boy repeatedly grabbed my butt during a slow song at a middle school dance even after I asked him to stop. A young employee who found my phone number when I signed up for a gym membership called incessantly till my dad complained to the gym. There were all the ones who were hurt or upset because I didn't like them the way they liked me, and it was my job to make them feel better. And I complied, because I was nice, because that's how I thought I was expected to behave.

I told myself I was lucky. None of my boyfriends ever hit me. None of them forced me into anything. I took birth control to help tame my teenage acne, so I didn't need to worry about pregnancy when I started having sex. I told myself that they would grow out of it, or that I would. I'd stop being scared, stop freezing, start standing up for myself. But it didn't get better as I got older. It got worse.

One man grabbed my breast and squeezed as I walked by him in broad daylight with a group of friends; he responded to my curses with laughter. Two others assessed me in a hotel breakfast room, loudly debating my gender before deciding, "It doesn't matter, I'd fuck it either way." Another man crept up behind me as I was entering my apartment building and growled, "I'm coming in after you." When I swung around, preparing for a fight, he was laughing at his prank. He lived in the building, too, he explained.

At some point I began looking for things to destroy. I took a pocketknife to a pillow I sewed in my middle school home ec class. The blade tore through the fabric with a satisfyingly jagged noise. In less than five minutes, the fabric was shredded, the stuffing mangled. I destroyed old clothing, shattered pottery my partner made that hadn't turned out right, burned scraps of paper. Anger lived on the edge of my vocal cords, a scream that built but never manifested, as if I were stuck in a nightmare, voiceless.

Sometimes I think that anger calcified in me, that it became the painful knots of tissue doctors would later excavate from the folds of organs within my abdomen. There's no good outlet for a girl who's angry at half the population. Only gods get to inflict their wrath wholescale, in world-altering events like floods.

We don't know exactly how our Earth earned its skin of water. Meteorites from elsewhere in the solar system may have been carrying water when they crashed into our proto-planet.[1] Or maybe the veil of hydrogen in the early atmosphere was a chemical catalyst, interacting with the churn of magma oceans to create water.[2] Whatever its source, we were born from a wet planet.

Water is always on the move, freezing and melting and falling and evaporating, pushing inexorably through sediment and stone, making the circuit between land and sky. Seas and lakes and rivers and glaciers, all storage basins that feed and receive the clouds. Without water, there is no life; without life, there is no reason to name a flood. Water has always followed its own whims. It wasn't until organisms moved beyond its embrace that life and water found themselves at more regular cross-purposes. What is a flood if not water going where we don't want it? And with that intransigence comes the need for control.

Human societies have engaged in water management for at least ten thousand years, and quite possibly longer; there's only so much the archaeological record can tell us.[3] The tangible evidence we have is incredible: abandoned wells on Cyprus from 8300 BCE; a dam in the Levant dating to 7500 BCE; a network of canals created by Mesopotamian civilizations after 3300 BCE in response to a period of arid weather.[4] Societies from ancient Greece to South America constructed massive aqueducts to shuttle water where it was needed. Scholar Giulio Boccaletti has even argued that the concept of a state appeared in large part because of the need to manage water as a resource.[5]

But human ingenuity can't account for every circumstance: the wind and the tides and the trees and the deserts and glaciers and volcanoes and deep-sea earthquakes all affect the movement of water. Even dams—well thought out by engineers—can come undone. And when disasters occur, a common response is to look for some force to appease or some person to blame.

The Babylonian creation myth *Enūma Eliš* describes the goddess Tiamat, who begins as the primordial sea but becomes an adversary of male gods when she creates an army of monsters.[6] Tiamat is ultimately defeated by Marduk, the patron god of Babylon, and he uses her body to create heaven and earth, with her eyes serving as the source of the Tigris and Euphrates. It was her death that created both life-giving rivers.

Outside the realm of myths, blame has been directed toward people rather than deities. Look at the centuries-long climatic shift in the North Atlantic known as the Little Ice Age. Lasting from the fourteenth century till the nineteenth, it caused several rounds of unusually cold temperatures across the Northern Hemisphere. Scientists have suggested multiple possible contributing factors: a period of reduced solar radiation, a change in ocean circulation, large volcanic eruptions, the genocide of Indigenous people in North and South America and the regrowth of forests where once they cultivated the land.[7] The climate is influenced by dozens of factors, sometimes impossible to prize apart. The exact reasons for the Little Ice Age remain unclear; the effects are more obvious.

From 1315 to 1317, around the start of this climatic disruption, northern Europe experienced the Great Famine, linked to a surplus of wet weather and severe winters.[8] Farmers struggled to plant crops in sodden fields. Diseases spread and killed livestock. Food production plummeted. It's thought that somewhere between 10 and 25 percent of Europe's population died, mainly from starvation.[9] Then, while people were still struggling with insufficient nourishment, the Black Plague arrived. And the weather did not improve.

Until the instability wrought by the Little Ice Age, Christian dogma held that humans had no power over weather; it was a reflection of God's wrath or contentment.[10] But as decades of harsh weather caused social instability, weather-making became the provenance of witches. Pamphlets described the appearance of witches in communities across Europe, including fantastical details about their trials and executions.[11] Prints showed wrinkled crones summoning rain showers. Older women were often the ones charged with witchcraft. They were seen either as dangerous for the types of work they performed in the community (such as healthcare) or as burdens who drained community resources.[12]

Once accused, the women were often tortured into giving confessions. Long needles pierced their flesh, thumb screws and leg screws clamped bones, straps bound their wrists behind their backs, from which they were lifted till the shoulders dislocated.[13] Then came the fire. Often multiple women were burned at once, on the rationale that the scale of their crimes could only be committed by a whole coven. It's hard to say how many people were killed in total, but historians estimate the number could be as high as fifty thousand.[14]

On the other side of the world, a similar display of power over the vulnerable asserted itself. Under the Incan Empire, one of the most important gods was Illapa, a weather deity responsible for snow, hail, storms, and rain—the last being a crucial necessity for crops.[15] To ensure stable weather conditions, to commemorate the death of an emperor, and to unify the empire, the Inca offered ritual human sacrifices known as *capacocha,* in which young children (often girls) were given drugs, then forced to make the long voyage up steep mountains, where they would be killed.[16] Whatever the precise motivation, the sacrificed children were thought to be messengers to the gods and had to be unblemished in every way—including being virgins.[17]

More than one hundred bodies of sacrificial victims have been recovered from mountaintops, where they were discovered to be naturally mummified through freezing and desiccation.[18] This was the case in 1995, when an anthropologist and a mountaineer hiking near the summit of a dormant Peruvian volcano found a body. Juanita, the Maiden of Ampato, they called her, after the 20,000-foot mountain peak that sheltered her in death. She was a young teenager when she died, most likely from being bludgeoned in the head. The anthropologist, Johan Reinhard, later wrote a book and reminisced about the encounter, "I couldn't take my eyes off the face of a mummy that seemed to be staring directly at *me*."[19] He helped transport the girl's body down the mountain, then ushered her through the many scientific tests that provided greater insight into her life and death. Genetic analyses, studies of the food in her frozen stomach, and finally, a reconstruction of her face in silicone.[20]

Photos of this model peppered news media beginning in October 2023. My first reaction was horror. I've never felt comfortable seeing mummies displayed in museums. Neither giddy excitement nor morbid curiosity feel appropriate when viewing their remains. How would they have responded to this odd afterlife, enshrined in an unimaginable world to be stared at by strangers? It seems disrespectful. But even worse is bringing a sacrificed girl back to some facsimile existence, letting the whole world consume her image after her own society took her life.

I understand the impulse to wrench a human loose from the passage of time. Haven't I done it often enough in writing about the past? I regularly wish for something more than a two-dimensional photo of a dead loved one. Looking at photos of Juanita's mummified remains versus her sculpted face are viscerally different experiences. All my emotions are stronger when I see her *likeness* instead of her physical body. It's not only horror. It's anger and sadness for her fate, fascination about her life, admiration at the skill of the artists who crafted her. Looking at a mummy is seeing someone dead. Looking at a model is imagining her alive.

Oscar Nilsson, the reconstructor, is famed for bringing ancient faces to life, from Neanderthals to Cro-Magnon. With a background in archaeology, he uses scientific techniques to create the most realistic replica possible. That means genetics research, and an understanding of physiology and markers of malnutrition, and measuring every facial feature. The one place he takes any liberties is the expression.

"In Juanita's case, I wanted her to look both scared and proud, and with a high sense of presence at the same time," Nilsson said in an interview.[21]

It's entirely possible that our modern facial expressions vary from how they appeared five centuries ago. Even between cultures, there are different ways of showing happiness, humor, pain. Nilsson is an expert in the unique combination of muscles that contract to convey any expression; he's surely studied the subject in far greater detail than I have. But when I look at Juanita's reimagined face, I don't see fear or pride. I see tightly restrained anger.

Engineering has surpassed divine rituals in our efforts to control water. We have forecasts for rain, dams for rivers, seawalls for oceans, sump pumps for basements. As of 2022, some three-quarters of the global population had a safe and reliable source of drinking water, a number that could be improved so that *all* people have clean water but still represents an impressive feat for eight billion humans.[22] My morning glass of water comes from a pitcher that I fill at the kitchen sink, one of five taps in our house.

Given this convenience, it's troublingly easy to fall for the idea that humans have mastered water. It's easy to believe the question of water management is more an arithmetic problem than a social one, something to be solved with better pumps and wells and dikes and dams.

But water's absence and excess are still common phenomena, despite (and sometimes because of) our planetary manipulations. Every "natural" disaster has unnatural consequences; those who are most disadvantaged before a flood are most likely to suffer during and after it.[23] Nature may be impartial, but the way we've built our cities, planned our governments, provided health and social services is far from impartial. Even if we're no longer sacrificing young girls to mountain gods or burning women as witches (though this still happens), it remains women who are most vulnerable during natural disasters.[24]

When Hurricane Katrina made landfall on the Gulf Coast in late August 2005, it caused a catastrophic failure of New Orleans's protective levees. The storm came first, yes, but it tugged a flood along with it. Women made up about 80 percent of the people left behind in New Orleans.[25] It wasn't for lack of warning or lack of wanting to leave. One-third of city residents didn't own a private car in which they could flee; others had no money for gas or motels.[26] Women are more likely to be the head of single-parent households and to care for family members with poor health or disabilities; an evacuation would require transporting more people with greater needs.

Things didn't improve in the chaotic aftermath of Katrina, when a region whose land mass is comparable to the United Kingdom's was declared a disaster area.[27] All but one of the city's domestic violence shelters was swept away in the storm, and domestic abuse typically increases in the wake of a disaster.[28] One study found that women reported a 35 percent increase in psychological victimization and a 98 percent increase in physical victimization after Katrina.[29] But after 300,000 houses were destroyed and over 1.8 million were damaged, many women had no option but to live with abusers.[30]

The negligent federal response to Katrina was epitomized by President George W. Bush continuing his vacation despite the disaster. Later, he was photographed peering down at the flood damage from the window of Air Force One. Pundits would eventually say it wasn't the fabricated excuses that Bush used to launch a war in Iraq that sunk his approval ratings; it was the response to Katrina.[31]

In the fall of 2008, I traveled to New Orleans with a small group of university students. I was taking a semester-long course centered on the city's history, its demographics, the racial inequities, and the disastrous government response to Katrina. Thousands of families were still living in temporary homes. Compounding the flood damage from Katrina were more recent storms, including Hurricane Gustav and Hurricane Ike, the latter of which hit further west along the Gulf Coast. Although the category 4 storm dropped to category 2 by the time it made landfall in the US, the winds remained so strong that Ike cut its way up the country and caused collapsed trees and power outages across my college campus in Ohio.[32] It felt like the storms and the water they carried from a heating ocean were knitting us together, making the distance between southern Ohio and southern Louisiana seem smaller than geography would suggest.

In addition to watching jazz and blues musicians and scarfing shrimp boils and beignets, my classmates and I helped gut houses in the town of Lafitte, just south of New Orleans. We ripped down wood paneling, then moldy drywall, then insulation, filling the air with dust and spores that stuck to our sweaty skin. We breathed through masks, and I tied a bandanna over my short hair to keep it out of my face. Fluffy debris itched my skin, drifting through moisture-thickened air, a humid soup of flood wreckage and fungal life.

This sodden mess had been where someone lived, where they watched TV and ate meals and slept and talked and played. Did the family now have a visceral terror of water pouring in through the front door, smashing windows, lifting furniture? Had they spent days, then weeks, feeling desperate for food and fresh water, seeing bodies in the streets?

How is such an experience to be conveyed to anyone who has not directly experienced it? Witnessing never feels adequate, and sometimes borders on voyeuristic. I spent five months learning an abbreviated history of New Orleans and Katrina from the comfort of a classroom; did I have the right to be angry on someone else's behalf? And yet anger is a rational reaction to injustice. As psychologist James R. Averill writes, "More than anything else, anger is an attribution of blame."[33] I was learning to see reasons for anger beyond my own treatment at the hand of others, to move outward to a moral anger against systems of inequality.[34] I blamed the Bush administration for underfunding FEMA in the runup to Katrina, and the agency itself for its failure to coordinate with other organizations and provide help where needed.[35]

A flood is more than water ending up where it isn't wanted; it's everything that comes after, the success or failure of our social systems to help those who have been affected. A flood reveals whose suffering is given priority, and whose becomes a statistic.

A few weeks after I returned from New Orleans, there came a new media eruption. An Iraqi journalist removed his shoes during a press conference and threw them at Bush, one after the other. The American president managed to duck quickly enough to avoid being hit. The journalist was tackled, beaten, and removed by security.[36]

The incident was one of the first viral memes I saw evolve in real time. A Flash game called "Sock and Awe" allowed participants to take aim with a virtual loafer and try to hit the president behind his desk. Comedian Will Ferrell, impersonating George W. Bush on *Saturday Night Live*, performed a segment in which shoes were thrown at him from the audience.

What didn't go viral was the story of Muntadhar al-Zaidi, the man who threw his shoes. He was sentenced to three years in prison, which was then reduced to one year.[37] After being released, he claimed to have endured torture while there.[38] Even so, after his release, he said he had no regrets about the incident. He needed to express his anger against a monstrous political system that would allow a foreign nation to invade his country, kill its civilians, and perpetuate government corruption. His shoes and his words were the only tools available to him.

New Orleanians must've been equally tempted to throw something at the president back in 2005, when he finally came to make a speech about the strength of the city.[39] (Why is "strength" always evoked when no support is offered, when survival and death are the only two options?) Maybe, despite the many differences between a war in another nation and a hurricane hitting their city, they saw some commonality in that instinct to express rage. Anger can be destructive, certainly, but when strategically deployed, it can energize social movements. When Bush came to visit the city for the ten-year anniversary of Katrina in 2015, a group of protestors gathered outside one of the schools he visited. One sign proclaimed, "George Bush still hates black people."[40]

Between December 2023 and January 2024, I was riveted by another meme going around social media, this time on TikTok. It was a music video featuring a woman's silky contralto and the opening line "Get in the water." A duet ensues between the woman and an off-screen man, gradually revealed to be Odysseus. Near the end of the song, the woman threatens to "make tidal waves so profound, both your wife and your son will drown" unless he gets in the water (where *he*, presumably, will be the victim of drowning).

The first few times I heard it, I couldn't figure out who the woman was meant to be. *The Odyssey* has been one of my favorite stories since childhood. Maybe this scene was Odysseus listening to the Sirens while lashed to the mast of his ship? Maybe it was the witch Circe trying to convince him to stay with her?

Then I found the source of the song, a musical called *EPIC*. Morgan Clae was singing the part of Poseidon. The ocean god is furious at Odysseus for blinding one of Poseidon's sons, the Cyclops Polyphemus. I never would've guessed the song was meant for a man; the lyrics fitted too perfectly with the perennial zeitgeist of #femininerage. Within a few months of being posted, Clae's audio had been used in tens of thousands of other videos: women in goddess makeup, women with dripping hair, women snarling in fury. We wanted to be as mighty as a god who threatens to drown mankind. We wanted powerful men to listen to us, and to account for their deeds, and we'd even threaten the life of another woman and child to see it happen.

Can I call it coincidence that I happened to be reading a deeply infuriating book called *Unwell Women: Misdiagnosis and Myth in a Man-Made World* just as I stumbled on those videos? Or is there always some invisible feedback loop running subtly between pop culture and academia? From the book I learned the stories of three enslaved women leased by James Marion Sims in the 1840s. Sims is often called the "father of gynecology" for his surgical treatment for vesicovaginal fistula—a tear between the bladder and vagina, caused at that time by prolonged labor during childbirth.[41] He performed dozens of surgeries on Anarcha, Betsey, and Lucy, all without any form of anesthesia, even though some anesthetics were beginning to be used for women when giving birth.[42] Some scholars denounce his treatment of the women, which amounted to medical torture.[43] Others claim the women were social pariahs who willingly went along with him in the hopes of a cure.[44] Whatever your interpretation, it's clear who held the power.

Author Elinor Cleghorn describes dozens of stories in this vein, all eliciting a kind of secondhand body horror, combined with fury over the inhumane treatment these people received *because they were women*.[45] Things were worse for Black women, poor women, trans women, Native women, but even the upper echelons faced brutal treatment. Some women displayed symptoms of known diseases, like lupus or multiple sclerosis or kidney disease, but instead of being treated for those illnesses, they were lobotomized; two doctors who specialized in lobotomies were performing

75 percent of them on women by 1942, even though more men than women were confined in asylums.[46]

Even when something seemingly benefited women, such as the advent of hormonal birth control, the process by which the drugs were introduced is disturbing. Impoverished Puerto Ricans were selected as the test subjects and were not told they were participating in a clinical trial—nor were their side effects treated seriously by the leaders of the trial.[47] Even when three women died during the trial, no one investigated the cause of their deaths.[48] An added bonus according to the racist, eugenicist researchers of the day was that the birth control (along with sterilization) would result in fewer Puerto Rican babies.[49]

Anger is a normal response to a perceived breaching of our boundaries. Reading so many stories of injustice could easily make a person self-immolate with rage, convinced there's nothing in human history that merits redemption. But I haven't sought these stories out in order to douse myself in kerosene and strike a match. I don't want to burn for the world. I seek other's anger because I want to feel less alone in my own. Because anger may spur us toward change. Because without the armor of other women's anger, it would be too painful to tell my own story.

Between age fourteen and age twenty-four, my period rarely felt like a punishment. Not when I bled through a tampon, a pad, my underwear, my jeans. Not when I switched to a reusable Diva Cup and grew accustomed to the sticky blood and mucus pooling over its edges onto my fingers. Not even when I had a ruptured ovarian cyst, an experience so painful it caused me to vomit repeatedly and sent me to the ER on suspicion of appendicitis.

For years, I'd regulated my cycle with the help of pills or a vaginal ring, a daily dose of etonogestrel and estradiol. But sometime around college, the mood swings started. Crying jags interspersed with a numbness so intense it felt like I was simultaneously being crushed by and dissipating beyond the boundaries of my body.

Days of intense irritation at everything and everyone, in which I was constantly reining in my impulse to yell. Fear and anxiety stealing sleep from me at night. It only happened one week out of the month, but it was a miserable week. Sometimes I worried I wouldn't be able to keep surviving them.

And then came my Hashimoto's diagnosis and mood swings that were so clearly tied to thyroid medication. I thought I'd found an answer. Balance my thyroid, and I'd be better. This turned out to be easier said than done, as one doctor after another dismissed my symptoms. What if my thyroid was ungovernable in part because of my birth control? A number of autoimmune diet books and online forums convinced me that the choking depression I felt before bleeding was tied to those artificial hormones. The memory of my teenage periods pre–birth control was sweetened by nostalgia, and I was desperate for relief. I became convinced that without these additional hormones, I'd find a healthier, happier version of myself. The only problem was that I still needed some form of birth control to avoid pregnancy. That left one option: a hormone-free copper IUD.

I made an appointment at a family planning clinic. I knew very little about the procedure, except that it was said to be painful. But the clinic told me I'd be fine if I took some Advil beforehand. Four days after I finished my last artificially mediated period, I took what I thought would be a long lunch break from work and went to get my Paragard IUD.

They called me back to the room and had me strip from the waist down. A man came in, introducing himself as a resident who'd be performing the procedure. Was this an ominous portent? I didn't want to think so; everyone has to start somewhere. But my nerves ratcheted up a half-step, my pulse thumping. It had been just over a year since my heart surgery.

The resident guided a speculum between my legs, the pressure on my vaginal canal pushing me open. He warned of a pinch because he'd be holding my cervix open. The sharp pain was more than a pinch, but it was nothing compared to the device he inserted into my uterus, a sounder to measure its diameter. The deep, wrenching pain was so intense that I shrieked involuntarily and dug my heels into the stirrups, trying to lever my body away.

He asked me to hold still. My uterus was retroverted (tilted more toward the spine than its typical location over the bladder); it was hard to get the right measurements. My heart was thundering as if I'd just sprinted a 5K. Every time he used that sounding tool to tap the edges of my uterus, I screamed. The noise was guttural, a sound I didn't know I could produce. After too many knifing jabs, the IUD breached the barrier of my cervix and found its new home.

But then the doctor came in to review the resident's work and announced the IUD wasn't inserted correctly. They'd need to remove it and try again.

My back was so sweaty I'd soaked through my shirt and turned the crinkly paper on the bed to pulp. I was nauseous, dizzy, shaking. The agonizing pain was gone, but my uterus was still rippling with cramps. The doctor informed me I could go ahead with a second attempt or come back another time.

The thought of undergoing a second round was almost unbearable. It would be so much easier to go back to the familiar hormones, mood swings and all. But I didn't want that. The past few years had wrung so much from me, between my Hashimoto's diagnosis and my heart surgery. I wanted this IUD for my body because I believed, in the long term, it would provide some respite. I told the doctors I could do it.

A nurse gave me two more Advil and said they'd wait thirty minutes for the pain relievers to start working. I was chilly from my damp shirt and exposed legs, draped with nothing but a flimsy sheet. I dreaded the doctor's return, while also longing for her to get it over with. When she finally came back, the heat of anxiety rushed over my skin.

I'll give her this—she was more efficient than the resident. But she was also more brutal with the sounding device. It thumped the inner walls of my uterus and I heard myself beg her, "Please, no more!" This, I thought, was how women tortured as witches confessed to any sin under the sun.

At last, the copper IUD was inserted, this time in the correct location. I could wait a full ten years before getting it replaced; maybe by then I'd have blacked this experience out of my memory. I wrapped up in my jeans and a winter jacket, still shaky-legged, dry-mouthed, sweat-dampened. It was a cold March day and I was free, finally, of synthetic estrogen and progesterone. Now my body would do what it needed to do, naturally.

Eight years later, a gynecologist who specializes in laparoscopic excision surgery for endometriosis will sigh when I tell him I have a copper IUD. "I hate those things," he'll say. "They always make endometriosis worse." At this point, I'll still be learning whether I can trust this doctor, uncertain if he's the right person to help me. But the news about my IUD will be a devastating blow. I'll wonder who I'm supposed to trust about what happens in my body if I can't place my faith in the authority of doctors, or the counsel of other women.

WHEREVER WATER GOES

Picture glaciers encroaching from the north, white and gray and blue monoliths that heave and grind like living mountains. Or, imagine them as slow-moving lungs, breathing on a geological timescale. They expand on the inhale, retract on the exhale, their icy breath felt across thousands of miles of landscape. The glaciers scrape rock and soil and trees in their passage; they build borders across continents; they play tag with the global climate. Imagine they mark the contours of your world, the same as city skyscrapers that now direct our eye heavenward and block the horizon. And then, one day, as the glacial lungs are exhaling away, they unleash a lake of glacial melt that has long been pent up by an ice dam.

The flood is so intense that it spews forth at speeds reaching eighty miles per hour.[1] Water gushes from the Cordilleran Ice Sheet and scours the terrain so forcefully that it digs valleys into smooth earth. Icebergs surf down, gouging more land. When the water slows, it forms temporary lakes up to four hundred feet deep.[2]

These outbursts from melting glacial ice happened dozens of times between thirteen thousand and fifteen thousand years ago, reshaping the landscape that would be called Montana, Idaho, Washington, and Oregon. Today we name them the Missoula Floods. Considering the accumulation of evidence indicating humans arrived in North America more than twenty thousand years ago, it's possible there were people who witnessed the cataclysm—or died in it.[3]

Floods are a natural phenomenon, shaping the planet's topography as dramatically as any volcano. The difference between floods for most of Earth's history and those that occur today is that we've changed the composition of water around the world. When we think of climate change, we look to fossil fuels clogging the atmosphere, increasing the planet's temperature and pushing seas higher. But that ignores all the earthbound elements we have poured into our waters, the innumerable chemicals that come not from leeched rocks or soil but from manufacturing.

The human impact on water quality stretches back to Antiquity. Ancient Romans caused the eutrophication of at least one lake that we know of when they plowed the land around it.[4] Even after Roman occupation ended, it took around three hundred years for the lake to recover. The Romans also contaminated river water with their use of lead pipes, a lesson we still haven't learned from.[5] The transformation of landscapes only expanded in the centuries after the fall of the Roman Empire. Population growth in the Middle Ages of Europe led to a proliferation of tanneries and textiles mills. There were around half a million mills interrupting the flow of rivers in Europe by the eighteenth century.[6] The byproducts of tanning and fabric production were dumped into the water, whether that was fermented urine or ash and lye. The byproducts of papermaking were also noxious, with pulp slurry regularly being dumped into rivers.

The Industrial Revolution accelerated the process and allowed people to value profit over visible changes to water and air.[7] Surely, with such massive quantities of both elements, our human endeavors would never manage to alter the composition of either. Whether that was toxic fumes released into the air, or industrial waste poured into the water, industry magnates made regular use of these two abundant elements, trusting them to dilute the contaminants to innocuous levels. Thanks to John Rockefeller's Standard Oil refineries, New York City's three petroleum districts were each producing three hundred thousand gallons of waste material and dumping it in the water every week during the 1880s.[8] The pollution was so devastating that fish, shellfish, and bird populations were all devastated. The water turned ink black. Citizens protested this disregard for human health and the environment for decades. Finally, in 1924, the Oil Pollution Act was passed—but the oil corporations had such influence in the government that they were able to convince politicians to revise the act so that onshore factories were exempted from any regulations, leaving only those at sea to comply.

This level of pollution was not unique to New York. Heavy industry led to the creation of hundreds of superfund sites, places where the owners neglected their responsibility to clean up dangerous pollution.[9] The Cuyahoga River, a tributary of Lake Erie, famously caught fire at least a dozen times because of steel mills, paint factories, and oil refineries.[10] At the time, the pollution was accepted as proof of Cleveland's successfulness, of good jobs and a booming economy.[11]

Although the river fires helped lead to the Clean Water Act, passed in 1972, that legislation itself has only been partly successful. A report in 2022 found that 50 percent of the rivers and streams assessed for pollution are still "impaired," meaning they aren't safe for swimming in, drinking from, or fishing out of—let alone being healthy for the wildlife that live there.[12] And this is in large part due to the government writing laws that favor industry, along with its failure to enforce the few prohibitions that can actually hold industry accountable. In the 1976 Toxic Substances Control Act, which was meant to give the EPA powers to protect humans and the environment from harmful industrial chemicals, sixty-two thousand chemicals already being used at the time of the law were exempted from inclusion.[13] Industries fought the EPA in court whenever the agency tried to enforce its laws, making it difficult to prevent chemicals from entering the environment. This sometimes means that water remains polluted for decades before any government agency is willing and able to recognize it as such.

Communities have experienced this reluctance to act with per- and polyfluoroalkyl substances (more commonly known as PFAS), chemicals that can create infertility, thyroid disease, increased cholesterol, an increased risk of kidney and testicular cancers, and cause immunotoxicity—making people more vulnerable to infectious diseases.[14] At least 45 percent of tap water across the United States is contaminated with one or more types of PFAS.[15] There are more than twelve thousand types, and not all have tests that can detect them.[16] They have been filling waterways for decades, but it was only in 2024 that the EPA created a rule labeling them as "hazardous substances."

When water overflows riverbanks, or pushes past high-tide boundaries, it isn't only the water's havoc that deserves our concern. It's also what is *in* that water—what we put there. Floods don't simply ruin furniture and set the conditions for mold in walls and floors. They pass on all the chemicals the water carries. Pollution hasn't disappeared just because we no longer have to look at it.

The floodwaters came from the lake, blown beyond their habitual boundary by relentless wind and storms. Centuries earlier, most of Lake Erie's western shoreline was wetlands, home to a riot of birds and fish and numerous Indigenous people: the Erie, Wyandotte, Kaskaskia, and Myaamia.[17] Today the marshes are mostly gone, replaced by houses and marinas and a coal-burning plant and nuclear plant and a wickedly good amusement park called Cedar Point that bills itself as "America's roller coast" for the view of Lake Erie that riders get when they're screaming at the top of a 310-foot hill. Because the absorbent buffer of wetlands only exists in about 5 percent of its historical range, when water overflows its banks, it rolls into all these human structures, both depositing and collecting chemicals that it will tug into its body when the wind recedes.[18]

We met the flood just before twilight, my partner and I, driving from Chicago back to our Ohio hometown. Family reports of the coming flood had trickled my way via social media, but I didn't expect a maze of country roads closed for the fifteen-mile stretch between Fremont and Port Clinton. Even the open roads shivered under a layer of black water, at times so deep we worried our low-slung Honda Civic would stall out. Weren't cars the most dangerous place to be in the event of a flood? But the air was mostly calm, the skies mostly clear, the lakeshore miles away. It seemed safe enough to ford these ephemeral streams.

On we drove, snaking across a landscape quilted by fields of corn and soybeans. It took twice the normal length of time to make our way to Catawba Island, and the flood only worsened as we approached. It was Memorial Day weekend and the carnival rides and tents that appeared annually for the city's Walleye Festival were inundated. The town's two main roads were covered in multiple feet of water; houses on the lake stood like soggy children wading in murky water up to their knees. Grassy yards had vanished.

We made it at last to my in-laws' house; they were spared from the flood by virtue of being on slightly higher, rocky ground. The city declared a state of emergency. Houses along the lakeshore suffered from flood damage, all from the waters that made their lakefront property so valuable. One might think that on these inland oceans, there is no sea-level rise to fret over. But the Great Lakes rise and fall based on snow and ice cover and precipitation and myriad other factors.[19] While warmer air temperatures have led to increased evaporation and reduced ice cover (meaning the water level decreases), heavy precipitation also increases with warmer air temperatures (meaning the water level increases). If researchers can say anything about the water level in the Great Lakes, it's that climate change is making them fluctuate dramatically from year to year.[20]

When I was growing up, my body extended to include Lake Erie. The velvety, viscous muck in the shallows, perfect for flinging cartoon-like tar balls at my cousins. The rainbow oil slicks of diesel on water around marinas that I wanted to touch but that smelled foul. The nets of algae wrapping around my legs when I swam, clinging to the centerboard and rudder of my little sailboat. The very water piped into our house came from the lake—cleaned and fluoridated, yes, but it was still Lake Erie water.

Every season, every shore had its own smell. The dusty freshness of snow and ice in winter. Pungent algal growth in spring, ripe as fertilizer on flower beds. Sticky summer thick with the fishy carcasses of mayflies, collecting in piles so thick that people bring out snowplows to push them away. Then, sweet autumn leaves blowing to the lake, taming its funk and exuberance.

Each beach had its own treasures: blue and green sea glass; the white lucky stones with their curved L-shape on them, otoliths from the inner ear of sheepshead fish; driftwood warped by its watery journey. Sometimes such a dense flock of seagulls promenaded on the sand that you could run screaming into them and feel your body disappear between white and gray and black feathers, human shrieks joining the shrill of the birds.

I caught perch and bluegill and smallmouth bass, I swallowed accidental mouthfuls of lake water, I watched the glittering curve of the Milky Way appear overhead on dark nights. I breathed in spray, washed my bloody scraped knees in water that held hundreds of microorganisms. When I first peered through a middle school microscope to look at the transparent body of *Daphnia galeata galeata*, a water flea, my view of the lake bloomed even wider. Lake Erie was alive, not just full of life. When I touched its waters, they touched me back.

For years now, I've wondered if the lake's story of pollution is tied to my story of illness. If we both got sick because we were stuffed full of microplastics; the level of them in the Great Lakes are comparable to those of the Pacific Garbage Patch.[21] If the lead and other heavy metals that concentrated in the flesh of fish was transferred to my own flesh, forcing my endocrine system and circulatory system and immune system to malfunction, uncertain of which molecules belonged in the body and which should be evicted.[22] I can marshal pieces of evidence for my theory, red-stringing studies together on a conspiracy-style murder board. A paper here that points to the link between endometriosis and endocrine disruptors, and those endocrine disruptors are present in the lake and negatively affecting fish.[23] Another paper there that suggests PFAS might cause thyroid disease.[24] I can look to all my aunts and uncles—my parents have three siblings each and all six of them have at least one autoimmune disease. They have also all spent large parts of their lives on Lake Erie.

These are ultimately futile explorations, I know. There is no way to unravel the mixture of environmental and genetic factors that lead to illness. And still I wish for definitive proof, a villain to point to, be that a corporation or factory farm or corrupt politician. I want a trial, an unveiling of autoimmune disease clusters that results in restitution for victims and a promise this won't happen again, similar to the lawsuits that have been won over cancer clusters. I want justice. Because my body has suffered so much, and the bodies of others have suffered so much, and we are told to accept that illness is simply part of life. This last may be true, but when the illness comes as a result of poisons, shouldn't we at least try to get rid of them rather than accepting it's the price to pay for modernity?

Before the thyroid diagnosis, before the heart problems, before the hint of irregular bleeding and pain, there was my bladder. My first UTI came at age eighteen, when I was working a summer job at the farm stand, spending hot days plucking fuzzy white strawberries from large crates, digging through baskets of potatoes to find the wet, rotten ones. I learned the specific stench of every kind of decaying produce, as well as all the gossip about the farmer's family and business.

It didn't start with pain, not at first. More of a pinch, a twinge, a pressure. An irritation that saw me visiting the restroom far more often than normal to pee. By evening, though, it wasn't only a spasming need to empty my already-empty bladder; the pain came, so intense it wiped out my ability to think. Like peeing fire, or shards of glass, with blood appearing in the toilet bowl as if to support that conclusion.

My mom was the one who explained I likely had a urinary tract infection. UTIs were apparently something of a family heirloom; both my mom and her mom regularly faced off with misplaced *E. coli*. She'd take me to the doctor in the morning, get some pain relievers and antibiotics, and I'd be good as new in a few days. A welcome reassurance, but not enough to dull the burning. I barely slept that night, spending instead long hours in the bathroom thinking that if I could expel every single drop of liquid, maybe the electrification of the nerves up my back and down my arms would finally dissipate.

So began my genetic inheritance: UTIs so frequent that one doctor suggested I take a prophylactic antibiotic every time I had sex, a suggestion I vetoed for fear of ending up with an antibiotic-resistant infection. I tried all the alternative cures: showering before sex, peeing after, drinking unsweetened cranberry juice, taking probiotic supplements that claimed to break up bacterial biofilms. The only thing that seemed to have any effect was D-mannose, a sugar related to glucose. I drank a slimily sweet glass of the stuff every few days, extra after sex.

Even with all my attempts to stave them off, the UTIs visited with infuriating frequency. I began traveling with Azo pain relievers. I bought pH strips and kits to test the amount of white blood cells in my urine, results I could pass off to urgent care doctors to expedite the antibiotic prescription process. But on several occasions, the nurses decided I'd need a bacterial culture to check what strain of bacteria had colonized my urethra to be sure I was prescribed the right antibiotic. On more than a few occasions, to our combined surprise, the results came back negative for *any* bacteria—but by then I'd already be halfway through the course of antibiotics and was generally told to finish them.

They told me some people were simply predisposed to UTIs; all other tests had ruled out yeast infections and STDs. The only solution was to wait for them to happen, treat them when they did, try to limit the amount of time I'd spend in pain. I felt I had no other options.

Here is the patient's dilemma: if you are a woman (especially a Black, Indigenous, trans, or otherwise minoritized woman), doctors are less likely to take your concerns seriously.[25] In my experience, the more emotive you become in expressing your needs, the less likely those needs are to be met. If you spend enough time in the medical system, you learn how to manipulate it. The right outfit to wear, the right amount of knowledge to display about your condition, the right way to express your pain. But you also learn that there are limits, *so many limits* to how we understand the body and can remedy it. By the time a doctor suggested I might have interstitial cystitis—chronic inflammation of the bladder with an unknown cause—I'd already come to that conclusion myself. Of course, there is no cure. But I'd been trying to manage this problem on my own for years, and it had only gotten worse. I decided, reluctantly, that I needed help.

I didn't mean to schedule two urology appointments in an eight-day span. My general policy is to put at least two weeks between new doctors because they can leave me feeling emotionally wrecked. But one doctor had to reschedule an earlier appointment and I didn't want to delay any longer, so here I was.

The first urologist was a woman; we instantly had a good rapport. But I was at her office for a cystoscopy, in which a camera would be shoved up my urethra to look at the inside of my bladder. Obviously it wasn't some Kodak monstrosity with a telephoto lens, but I got a peek at the device before lying on the examination table and it was about the width of a marker, which was going to be plenty uncomfortable.

A news report about the location for the new Chicago Bears stadium was playing while the PA prepped me, filling my urethra with lidocaine, wiping me down with iodine. She commented on her own hopes for where the stadium will be. I told her I don't watch football.

The doctor came in and muted the TV, adjusting the chair so I was fully horizontal, legs splayed by the stirrups. I gripped my fingers into sweaty fists, praying this wouldn't be a repeat of the IUD debacle. The doctor assured me it wouldn't be, but I'd seen the size of that cystoscope.

Surprisingly, the insertion wasn't too bad. Painful, but not excruciating. The doctor informed me she was filling my bladder with water to get a better look. This I barely felt. She murmured a few comments to the PA, then retracted the camera and removed the cystoscope. Another burst of pain, but then it was done, having taken all of five minutes.

"You have small patches of inflammation on your bladder," she said. "Hunner's lesions. I recommend we do a procedure where we put steroid shots directly into the bladder wall, maybe remove the inflammation. We'll do it under anesthesia so you're not totally traumatized. Some people have relief afterwards for six or eight months, even a year. For some people, it does nothing."

So I'd potentially have to get my bladder stabbed multiple times, even in the best-case scenario. It wasn't an encouraging thought. "And if it doesn't work?"

"There are medications we can try," she said. "But I think we should start with this."

After a few more questions, I was ready to use the bathroom and go home. No particular emotion arose at the confirmation of this new diagnosis and treatment plan. Not anger or shock or sadness. I was sure those feelings would come; another thing I've learned is that I tend to go through a period of numbness before the new reality asserts itself.

But the numbness that wore off fastest wasn't emotional; it was physical. Urinating came with a sharp, excruciating sting. And I was peeing blood. A lot of bright red blood. Whereas a UTI feels like peeing fragments of glass, this was like I'd been cut by glass and now had to pee on that wound. Maybe that sounds like a meaningless distinction, but believe me when I say there are infinite gradations in the typology of pain, and that these pains are different. And I didn't want to learn what it would feel like to pee after parts of my bladder wall had been cauterized.

I decided to keep my appointment with the second urologist. I wanted another opinion.

This urologist was part of a hospital system rather than a private practice and the differences were immediately apparent. More staff, a bigger office, clear procedures to follow: provide a urine sample, go to a room for height and weight measurements, get a blood-pressure reading. Then it was time to get undressed below the waist and hop onto the exam table.

I didn't feel any immediate connection with this doctor when he came in, but he listened to my history and asked a few questions. He brought a young woman in, a resident closer to my age than to his, and he offered occasional asides to her. I mentioned my history of endometriosis and he wanted to know what stage I was diagnosed with. When I told him it was stage 4, the worst, he turned to the resident and said, "Sometimes doctors will say it's bad, but then we go in and look at the photos and it's really not."

The same thing happened when I told him about the cystoscopy results. "Hunner's lesions are very rare in someone like you," he said. "How old are you?"

"Thirty-four."

"Hmm." Again he turned to the resident to say, "We don't do a cystoscopy for painful bladder anymore. Usually there's nothing to see."

I agreed to let the resident be part of this consultation because I value firsthand experience in education, but it was feeling more and more like the doctor thought of me as some Anatomical Venus, a wax model of a woman whose skin can be pulled off to reveal the organs below. They were used in medical schools and museums of the eighteenth and nineteenth centuries, beautiful and grotesque, uncanny in their verisimilitude.[26] They could not speak for themselves.

"Would you recommend a fulguration like the other doctor?" I asked, because that's what I'm really here for.

"No, I wouldn't go straight to the OR. There's something we can do in-office today if you want. It's called a bladder instillation." He explained how the combination of medications work: heparin is thought to help patch up the mucus layer of the bladder, while lidocaine provides pain relief. The procedure would be mostly painless; the tube used is even smaller than a catheter—it's an infant feeding tube. It sounded better than surgery; I agreed to this treatment, even though I'd have to do it weekly for a month.

Once again I was on my back in stirrups, legs splayed for more strangers. The doctor slid the tube up my urethra and I hissed with pain. It was less intense than the cystoscopy, but worse than a needle entering a vein to extract blood.

The doctor shifted away from me and said to the resident, "You'll see this in patients with more severe cases of interstitial cystitis, where even this small of a tube causes irritation."

My annoyance flared into anger. I had a sudden urge to scream for him to stop. Maybe that's why, when he flushed the first liquid into my bladder, a slight burning was followed by an intense wave of nausea, so visceral it was hard to keep myself from pulling away. Both sensations passed quickly; I breathed, and thought about the fantasy book I was reading, my current escape from my body. In another minute all the liquids were pumped in. When the tube was removed I hissed again. The resident smiled at me and said, "It's out now."

I tried not to snap at her. "I could tell."

The instillation had to stay in my bladder for at least thirty minutes. I was instructed to lie first on one side, then the other, so the liquid could evenly coat my bladder. I read my fantasy book. I tried not to think of anything else. When it was time to go home, I peed. No blood this time. Still pain. And no emotional numbness to protect me anymore.

I felt flayed by the ten thousand tiny cuts doctors inflict on their patients. When I recounted my experience to a friend, she said, "I know you're so generous, but you're already going through a lot.

You can say 'no' to a resident coming in. It's okay to take care of yourself."

I knew she was right, but having a resident in the room was something of a litmus test. It let me see straightaway the difference in the doctor's interactions with his students versus those with his patients. It was a shortcut to learning whether I could trust the doctor, even if it's fucking painful. And now I knew that if this treatment didn't work, I'd be going back to the first doctor for surgery.

A Greek physician named Hippocrates treated women suffering from the symptoms of wandering wombs. Their rogue uteruses caused everything from a sense of suffocation to headaches. When we talk about the history of medical misogyny, we point to Hippocrates and his catch-all diagnosis of hysteria that has plagued womankind since Antiquity.[27] We blame him for how we're still treated several millennia later.

Except that history is not entirely true. The term *hystera* refers to maladies related to the womb.[28] It's true that the Ancient Greeks *did* believe the uterus moved around the body, but this was just one small part of the afflictions that could arise from this organ. Perhaps more important than the uterus's movement was the regularity of menstruation. The Greeks believed women absorbed more fluid than men, and if they didn't bleed once a month, the surplus blood would build up in her body and cause illness.[29]

Hysteria as a catch-all diagnosis for any illness (or any behavior deemed inappropriate) was propagated much later, when French philosopher Emile Littré translated Hippocrates's texts into French and retroactively applied the term *hystérie* as a header to several sections and as a diagnostic term.[30] This iteration was not something that appeared in Antiquity. In the nineteenth century, medical doctors were still establishing themselves as valid professionals with exclusive knowledge. Creating a link between their treatments and the Hippocratic texts was a way of creating a tradition.[31] They hearkened back to ancient Greece because it gave more weight to their assertions. When Fleetwood Churchill, author of a series of monographs in the mid-nineteenth century on diseases facing women, described a twenty-one-year-old woman who was bleeding from the ears and vomiting blood, he followed Greek tradition and attributed her maladies to a missed period.[32]

The story of Isaac Baker Brown is equally illustrative. The surgeon established a practice with a mouthful of a name: the London Surgical Home for the Reception of Gentlewomen and Females of Respectability suffering from Curable Surgical Diseases.[33] Some of his work was truly innovative, including surgeries to treat ovarian cysts and repair vaginal tears. But for patients suffering from hysteria, idiocy, mania, urinary incontinence, and other ailments, Brown regularly performed a clitoridectomy—the removal of the clitoris. Sometimes women were given clitoridectomies because they sought a divorce from their husband; after the operation, they were returned to that same husband they wanted to leave. Problem solved. Brown later defended his use of the clitoridectomy by claiming it was a procedure that dated back to Antiquity.[34]

We do have examples of women who wrote their own treatises on health, who distributed remedies, who oversaw the childbirth process. But those stories have not been transmitted with anything like the regularity of men's stories; they weren't the examples men drew upon when they established the medical profession. Anything that gave women autonomy was seen as a threat, something to be suppressed rather than encouraged.[35] It has taken concerted effort to reestablish the long tradition of women healing and teaching one another. And even after generations of clawing back bodily autonomy, the right to tell our own stories and choose what happens to our bodies, those gains are still unequally distributed, and so vulnerable to attacks.

The medical profession's approach to pregnancy is the archetype for all the scrutiny, judgment, and mistreatment women face. Black women have the highest maternal mortality rates of any demographic in the US; they're at least three times more likely to die during pregnancy than white women.[36] This inequity has many causes, from medical bias to institutional racism to preexisting conditions before pregnancy. But undoubtedly its roots lie in the country's history of chattel slavery, when Black women were valuable not only for their field and house labor, but for the labor of childbirth, because their children would also be enslaved, no matter the race of their father or the circumstances of their conception.[37] Enslavers closely monitored enslaved women's fertility and the childbirth process, blaming the women themselves for infant deaths.

In the aftermath of the Supreme Court overturning *Roe v. Wade* in the summer of 2022, surveillance of women's bodies increased. Black women, poor women, any group of women that has some minoritized identity is more likely to suffer under these reproductive restrictions—and to be targeted for the charge of feticide if they have a miscarriage or any other medical challenges during pregnancy that result in the loss of the baby.[38] This was the tragic case of a Black woman from Ohio whose hospital failed to answer her questions about her pregnancy. After she left the hospital, she miscarried at home. Police then arrested her and charged her with abuse of a corpse.[39] The charges against her were ultimately dropped, but more states are enacting harsher punishments for those seeking abortion (including proposed "homicide" bills that would allow patients seeking abortion to be charged with murder).[40] The result of this increased obsession with pregnancy is that more people will be imprisoned for wanting to choose what happens to their bodies.[41]

Reproductive rights are intimately linked to health writ large, and it can be dangerous for some people with uteruses to be pregnant due to certain health conditions. As legal scholar Michele Goodwin writes, "Robust legislating that chips away at reproductive rights and encroaches on women's reproductive healthcare is about more than abortion. Rather, it is about a fundamental lack of respect for the humanity, dignity, and citizenship of girls and women."[42]

It strikes me as particularly insidious that at the same time that women's rights are being eroded, multiple states have enacted laws limiting the kind of history that public schools can teach.[43] By forcing teachers to avoid talking about "divisive topics" like racism and gender identity, we're creating an unreal version of history in which the United States becomes a utopian ideal of a country, rather than what it is: a complicated, flawed nation built on Indigenous genocide and displacement, with the labor of enslaved Africans.

The erasure of history and the removal of women's rights is a human, legislative process. But we can just as easily employ water to do the job for us when it comes to getting rid of people and land we'd rather not see. More than ninety-two thousand dams of at least six feet in height have been built around the United States, and large dam construction in particular has had an outsized impact on flood valleys and watersheds—especially for Native reservations.[44] Over 1.13 million acres of tribal land have been flooded by dams, often without the consent of the tribes whose land was affected.[45] Just one example is the Grand Coulee Dam on the Columbia River, which displaced 2,250 tribal members from the Colville and Spokane Reservations and devasted the local salmon population, which the tribes relied on for food and cultural practices.

In these cases, the water changed its course because of human intervention, but that doesn't mean water always follows our rules. Manipulating water into the form we prefer can only ever be a temporary measure. On a planet where 71 percent of the surface is covered in water, we land-dwellers delude ourselves if we think we're in charge of the path that water takes. What will shake us from this self-induced stupor, the egotism that leads us to believe floods can be managed, mitigated, outrun? If we can't expect the government to save us, or the medical system, where do we turn?

FLOOD MEMORY

What came first: the rivers or the rains?

Our planet's water is trapped in a cycle, moving between land and sky, sometimes in solid form, sometimes liquid, sometimes gas. About 97 percent of the water resides in the oceans, leaving a scant percentage for the glaciers, rivers, streams, lakes, and the atmosphere itself. But that minuscule proportion of water living as vapor plays an outsized role in agitating its liquid and solid brethren, because water vapor is a powerful greenhouse gas. Without it, so little heat would be trapped near Earth's surface that it would barely be a habitable planet.[1] For every degree Celsius that our planet warms, the atmosphere can hold 7 percent[2] more water vapor, which creates more rain that makes rivers roll beyond their banks. As climate scientist Giulio Boccaletti writes, "Earth's climate is sensitive because there is water in it."[3]

Greater quantities of vapor not only create fiercer storms but also extreme levels of precipitation.[4] In 2022, floods in Pakistan coincided with heatwaves in Europe and China, leading to the deaths of more than 1,700 people and the displacement of over 30 million.[5] In September 2023, eight inches of rain fell on New York City across the span of a day, causing severe flooding and requiring the rescue of twenty-eight people trapped in the floodwaters.[6] In October 2024, parts of Spain received more rain in eight hours than in the previous *twenty months*, leading to hundreds of deaths.[7]

What came first: the hormones or my emotions?

Researchers have struggled to find the connection between menstrual hormones and mood.[8] A meta-review of forty-seven studies found no clear evidence for negative moods during the premenstrual phase.[9] And yet women are twice as likely as men to suffer mood disorders, like depression—a difference that begins in puberty and disappears after women experience menopause.[10] Another study of women in Denmark found that the use of hormonal contraceptives was associated with the subsequent use of antidepressants and depression, especially when contraceptives were started during adolescence—exactly the timeframe in which I started taking hormonal birth control.[11] I don't want to play into the misogynistic social myth that women become unhinged before their period, but I also don't know how to otherwise explain the way that my body feels flooded by depression and anxiety in cycles that follow the ebb and flow of my menstruation. Family and friends say I've always been sensitive: Is it that I'm more sensitive to the hormones at work in my body, or that I'm sensitive to *everything*? To the birth control, the objectification of my body, the scorn of doctors, the pain of misogyny, the chemicals that now permeate our planet?

Easier than quantifying my own sensitivity is measuring the planet's hydrologic sensitivity. No matter how complicated these calculations become, they are still based on external, measurable data. Quantifying a human's level of sensitivity verges on impossible because so much rests on the subjective experience a person has of their body. Is it more uncomfortable to walk along a busy road with the sounds and smells of traffic, or to stand in front of an audience and give a lecture? Does the emotional pain of a high school breakup come close to the intensity of emotions involved in a divorce? Why do I feel more immediate pain with a stubbed toe than a broken one?

I often wish that my body's processes were more quantifiable, so that I might have clearer answers and better medical treatment. As if the accumulated evidence of scientifically procured data might finally overturn centuries of medical misogyny. But that's playing into the demands of a rigged system, one that won't even think about listening to a patient's experience unless they have some biocertification to back it up, be it abnormal blood tests or scans.[12] And what good is data anyway, if doctors and health insurance companies deny us access to treatment? We have plenty of evidence showing our fossil fuel use is driving climate change, but that hasn't resulted in significant action on the part of governments or corporations.

The water cycle responds to human activity (among other things), but it is not concerned with us in particular. We're just the most recent inhabitants of a watery planet, one species among millions that have lived and gone extinct according to biological, geological, and meteorological dynamics. We've done our best to capture and control water, to bend its movement to our needs, but human history is awash with examples of water thwarting our best efforts. Research improves our understanding of hurricanes and flash floods and tsunamis, but it can't stop them from happening.

Data alone won't save us; emotion and memory deserve attention as well. Rather than pathologizing sensitivity, we should cultivate it—become more aware of the changes we're seeing and the emotions we feel in response to those changes. Feelings in themselves are not enough, but they can be a starting point for action.

I recently saw a video from the 2011 tsunami that hit Japan, in which a man at the top of a building is shouting from a bullhorn at people still standing along the seawall in Miyako. They're watching the uncanny appearance of water being pulled out to sea. Next, of course, it will come crashing back. The man and others are shouting to the people on the street that they need to get higher, that they need to run, that this is dangerous. Moments later, a roll of water moves through the inlet. Then black waves pour over the seawall, sending boats and cars tumbling as if they'd been caught by enormous rapids.

The wave wasn't what I expected based on Hollywood depictions of fifty-foot-high monsters suddenly crushing towns. It was still horrifying. But what I went back to, again and again, was the man with the red bullhorn trying to get people to understand their lives were in danger. What made him more sensitive to that danger? Knowledge, memory, experience, perspective? How many tsunamis had he lived through? Or was this his first, and it was only that he knew the stories because he'd heard them from other people in his community?

The question of what humans choose to remember or forget has been tackled by Professor Lindsey McEwen, a scholar of geography and environmental management who has researched the concept of "flood memory." She lives in Gloucestershire, which experienced severe flooding in 2007 when heavy precipitation caused the River Severn to overflow its banks.[13] The catastrophe resulted in some 350,000 homes being left without drinking water; 50,000 homes having no electricity; and multiple villages being completely covered in feet of water for weeks.[14]

"The big abbey in Tewkesbury is quite an iconic building that's in the center of town. It's been there since medieval times, and they wanted to have some sort of remembering a year on after this flood," McEwen told me over a video call. "There were big tensions in the town as to whether people wanted to remember or didn't, because of the trauma and that side of things, but others wanted to celebrate the community side of things."[15]

Is it possible to do both? Celebrate survival and acknowledge the trauma? Aboriginal groups around different parts of coastal Australia, whose presence on the continent dates back at least fifty thousand years, have been sharing their stories for millennia.[16] When researchers spoke with communities at twenty-one different locations, they found that the people's stories about a massive flood occurring some seven thousand years earlier closely matched the geological record of sea-level rise that followed the last Ice Age.[17] Their stories gave specific examples: a barrier reef that had marked the start of the coastline till floodwaters devoured the shore; an island that was once part of the mainland. Those details map onto what geologists have seen in the stones. The Aboriginals' oral traditions were so durable and accurate that they connected communities through time and space.

Imagine what our own culture and society might preserve and learn if we maintained such a strong connection to place and the environment. If we braided the stories of individuals into a larger tapestry instead of silencing some and prioritizing others.

Recently, at a community workshop on climate change that McEwen was involved in, nearly all the women wanted to reflect on their experience of the flood, more than fifteen years after the fact. They wanted to tell stories of helping others and remembering those who died, because it felt important. Maybe those women will tell the stories to their relatives, and the next generation will carry them forward in time. It will take practice, and reinforcement, and communities placing an increased cultural value on the transmission of memory from one generation to the next. But if such memories can lead to greater flood resiliency, it's worth the effort of telling and retelling them. As the atmosphere collects more H_2O and CO_2, the Earth will pull more heat to its breast, swaddling us all in downy warmth. Glaciers will melt and sea levels will rise. Storms will grow fiercer. More floods are coming.[18]

One of the earliest scenes in the movie *Portrait of a Lady on Fire* shows a young woman on a rowboat, clutching a large wooden crate as waves jostle the boat up and down. When the crate goes in the water, the dark-haired young woman dives in after it, despite her heavy dress. She must save it because that crate holds her canvases, the crucial supplies she needs for her work as an artist.

The movie tells the story of two women in eighteenth-century France, a painter and an aristocrat, who have a short love affair before the aristocrat goes off to be married to some nameless nobleman. The evolution of their affair dominates the story, but all around it are signs of women living their lives: a group of girls taking painting lessons, village women singing around a fire, the two main characters and the household maid playing cards. At one point the lovers accompany the maid to an abortion. Afterward, the aristocrat insists that the artist sketch the scene of the abortion. She poses as the healer, while the housemaid lies on her back, in the same position she took for the actual procedure. The artist begins sketching as they hold the pose, recreating the scene as it appeared earlier that day.

Portrait of a Lady on Fire is the story of women trying to find agency in a society that denies it to them, but more than that, they're women who want to find a way to tell their own stories, to pass them on. And, despite its title, it is a movie where water dominates. Both the main characters are immersed in the ocean at different points. The sea offers freedoms that can't be found on land; water is the element of excess and the subliminal. Alone on an island, for a short time unaffected by the judgment of men, these women can give themselves up to desire and love and grief and anger. No one can tell them what they feel is too much.

When I decided to go off hormonal birth control and switch to the copper IUD, I wanted to blunt all the painful emotions my body put me through every month. And they were extraordinarily painful: the fever of anxiety, the bleakness of depression. But I'd be lying if I said it was only the extreme emotions I wanted to avoid. I didn't want anything, not the regular sadnesses or frustrations, not even the excitement so effervescent it alone made me drunk. I wanted to be even-keeled rather than restless, tough instead of weepy, devil-may-care, not a worrier. I wanted to be anything but sensitive.

I didn't realize it at the time, but in trying to sever my connection from all emotions by rejecting hormonal birth control, I was attempting the same thing as those doctors who gave so many women lobotomies. They weren't trying to solve the very real issues women suffered from; they were simply trying to disconnect women's minds from their emotional experience of distress.[19]

This isn't to say we should ignore women's experience with mental illness; I am entirely in favor of treatments for depression, anxiety, bipolar disorder, and the many other illnesses that can manifest in the mind. When my extreme premenstrual mood swings failed to improve after stopping hormonal birth control, a gynecologist helped me find the right low-dose antidepressant to take in the week leading up to my period, which hugely improved my symptoms.

What I'm saying is that when we pathologize *all* emotions coming from women—when we police women's anger, or laugh at women's fear, or minimize women's pain—we are upholding that millennia-old framework of sexism and misogyny. And often enough, we're doing it to ourselves, because that's what we've been taught.

To unlearn a lifetime of these lessons, I've had to look to other women, our history and our memories. Like tracing the shape of water that eroded canyon walls, I've searched for evidence of how women let themselves feel and how they grappled with feelings about their bodies. It's an activity I'm not alone in. In a call with Vanessa Zoltan, podcast host and author of *Praying with Jane Eyre*, she described her experience of getting diagnosed with and treated for endometriosis as a community undertaking. Her friend was the first person to suggest her symptoms pointed to endometriosis. Her mentor confirmed that Zoltan was indeed sick and wanted to know what her plan was for finding help. But it isn't only the women living around her that have given Zoltan a sense of connection; it's women throughout history.

"There is something about feeling connected through illness. I wonder all the time about women who died. Mary Wollstonecraft died [after giving birth] because they didn't remember to remove all the placenta," Zoltan said.[20] How many others suffered from the effects of endometriosis?

Both she and I had read Hilary Mantel's memoir, *Giving Up the Ghost*, in which the lauded British author details her upbringing and her experience of being ignored, institutionalized, and ultimately given a brutal surgical treatment for her endometriosis. We raved about the book just as we mourned Mantel's experience. To live for decades with a disease that's mangling your insides—to write prize-winning novels in the aftermath of horrific surgery—is it any wonder she describes accumulating "an anger that would rip a roof off"?[21] After my call with Zoltan, I went back to Mantel's book and came across this passage:

> Nowadays, more than twenty years on from my trip to St. George's Hospital, everything about me—my physiology, my psychology, feels constantly under assault: I am a shabby old building in an area of heavy shelling, which the inhabitants have vacated years ago.
>
> I am not writing to solicit any special sympathy. People survive much worse and never put pen to paper. I am writing in order to take charge of the story of my childhood and my childlessness; and in order to locate myself, if not within a body, then in the narrow space between one letter and the next, between the lines where the ghosts of meaning are. . . . I have been so mauled by medical procedures, so sabotaged and made over, so thin and so fat, that sometimes I feel that each morning it is necessary to write myself into being—even if the writing is aimless doodling that no one will ever read, or the diary that no one can see till I'm dead. When you have committed enough words to paper you feel you have a spine stiff enough to stand up in the wind. But when you stop writing you find that's all you are, a spine, a row of rattling vertebrae, dried out like an old quill pen.[22]

No, I want to tell Hilary, *you are more than a row of rattling vertebrae. You are reciting the words to a play I've lived, you are a person who suffered and survived, you are passing on a story that has meant everything to me.* But I can't tell her anything, not in person or via email, because she died of a stroke at age seventy in September 2022, one month after my own operation for endometriosis. I can only thank her ghost, send my gratitude for telling a story that helps me feel linked to a greater history than my own. For making it easier to summon my own words in this impossible task of reconstructing who I am after a decade of doctors cutting and jabbing and adjusting the organs of my body.

Pain is an unreliable narrator. You cannot always point to its source, even with the best scanning technology. You cannot quantify it, because it is solely the experience of one person. It's unclear if you can accurately remember what it felt like once it's gone (other than the knowledge that it was bad and you'd rather not feel it again, thanks).[23] Its severity doesn't even correspond with the severity of tissue damage that's occurred.[24] And if you can't measure the sensation in any way, how exactly are you supposed to pass on a cohesive narrative, build a tradition of seeing something invisible?

Part of the problem is that pain is a biopsychosocial phenomenon. Processes in the body, as well as our moods and emotions, as well as social status and culture all contribute to the experience of pain. You can't untangle any of these three elements from one another because they're so tightly enmeshed. It's why spraining your ankle during a kickball game your team is losing feels more painful than breaking your big toe during a fun volleyball practice. (To say nothing of the embarrassment of crying in front of your classmates with the first injury and learning it's better to downplay these things in team sports.)

A lifetime of being told women were prone to exaggeration led me to silence pain whenever possible. Not only did I try to forget it as soon as it was past; I learned to mistrust it. If I can still walk, the pain isn't that bad. If I can still breathe, the pain isn't that bad. If I'm still alive, the pain isn't that bad.

Here I am on the couch, curled up around an abdomen so painful I can't talk. Here I am pulled over on the road because the pain was so intense I was scared to be driving. Here I am whimpering in bed, my partner suggesting I go to the hospital. Only when an on-call doctor confirms that abdominal pain strong enough to wake you up merits emergency treatment do I agree to go. The ER doctors admit me, administer morphine that only slightly touches the pain, send me for a CT scan. But because the scan shows nothing, they won't take further action. I'm returned home with no answers.

I'm so good at ignoring the pain, downplaying it, that when I first learn about endometriosis, I dismiss the possibility of having it out of hand. Women describe vomiting repeatedly, passing out, bleeding for weeks at such incredible quantities that they become anemic. The endometrial-like tissue can glue organs together, making every movement agony, like a torso filled with barbed wire.

I'm certain I don't have the right symptoms. I don't bleed heavily enough for one thing. My pain doesn't only happen during my period. I only *sometimes* have pain during sex. Bowel movements only *sometimes* make me cry. And I have no idea about my ability to conceive because I've never tried and I'm not in a hurry.

At what point did *I* become the unreliable narrator? I denied pain's presence for so long that it took more time and energy to convince myself that I needed treatment than it did to convince the gynecologist who diagnosed me. For years, I felt it within my body: sometimes the gnawing of small rodent, the clawing of a trapped bird; sometimes sharp, electric bolts; sometimes a nauseating squeeze, like a hand trying to rip my intestines from my body. But it wasn't constant, and when the pain disappeared, I made myself forget it. I told myself the pain didn't matter; my society showed me women's pain isn't real. I separated my body from my self, the perfect Cartesian model of a mind living fully apart from its flesh. Maybe in a different world, one where the memory of pain is more readily acknowledged and shared, I would have acted sooner. Maybe I'd have been unafraid to bring my symptoms to a doctor, and the doctor would have immediately treated me.

That's not the world we live in. In this world, I danced around doctors for close to a decade, hearing that maybe I had polycystic ovary syndrome (PCOS), or maybe I had premenstrual dysphoric disorder (PMDD), or maybe it was related to my thyroid problem. It took years for me to press the issue, to admit my pain merited attention. By the time I sought treatment, the disease had spread enough that it took nearly four hours of surgery to remove all the growths, rather than the two hours the doctor had predicted. The spiderweb tissue spackled my colon, my bladder, my uterus, my ovaries, my ligaments, my ureter. If I'd waited even longer, would the disease have blocked off one of my kidneys or penetrated my colon?

No one can answer those questions because I opted for surgery before it got that far. But speculation is also impossible because doctors don't know enough about the disease process of endometriosis.[25] The medical establishment has known about this disease for a century, but they're still guessing about almost every aspect of it, even what *causes* it.[26] And because surgery is the only way to diagnose it (and one of the best treatments), doctors sometimes fail to recognize the symptoms for years, especially if the patient is Black or nonbinary or a trans man.[27]

In many ways, I got lucky. My MRI showed enough evidence of endometriosis that doctors agreed surgery was appropriate. I had a surgeon who specialized in excision and listened to my symptoms. The laparoscopy was the worst surgery I've ever undergone but also the best. When I woke up I was so nauseated that I threw up everything that went into my stomach: ice cubes, water, ginger ale, apple juice, and Jello. Either my nurses or my mom, depending on who was in the room at the time, came to my side to apply counterpressure to the three newly made incisions on my abdomen as I heaved. I puked into round plastic bags with orange rims so that the liquids exiting my body could be measured. No matter what antiemetic medications I took, my vomiting remained so unmanageable that I was admitted to the hospital overnight rather than being discharged post-surgery according to the original plan.

But the surgeon who performed my operation not only visited to check on me during the day, despite having other patients to attend to, he also gave me a color copy of the photos taken during my procedure and made a video on my phone describing the procedure. He assured me they'd cut out every spot of endometriosis they could find and that I should have very good results once

I'd recovered. The nurses checked in on me multiple times in the weeks that followed, offering advice for recovery and what to do about lingering nausea. I've never been so miserable and felt so cared for by my medical team. They made me feel as if I had the support of a community.

Lately, when I think about how to shift the organization of power away from a corporate-political class, I think of community. I think of the women advocating for better healthcare, like Jenneh Rishe, a nurse and the founder of the Endometriosis Coalition. I first considered the possibility I might have endometriosis because of her Instagram account describing her health ordeals. She manages multiple conditions, some of which have required hospitalization, but her commitment to endometriosis advocacy is so strong that she gave a virtual presentation for the National Institutes of Health on the need for more endometriosis research funding even as she was struggling with symptoms from other diseases. I think of Zoe Mckenzie, a physiotherapist who offers classes catered specifically to people with chronic illnesses and disabilities. I think of Cheryl Crow, an occupational therapist who organized an educational program for patients with rheumatic diseases and moderates a support group for those who have taken the class. I think of the long history of disability activists who have screamed and fought with their bodies and voices to make the world more accessible for everyone.[28]

These communities of care can and do extend to the environment. Tish O'Dell, the consulting director for the Community Environmental Legal Defense Fund (CELDF), has spent her whole life around the Cleveland area, just ten miles from Lake Erie. She's always been enamored of the lake, but only recently came to think more deeply about the lake being within her. Not metaphorically, but literally, through the water she drinks and the air she breathes and the food she eats.

"What we do to nature, we do to ourselves, because we are nature," she tells me in a video call. O'Dell first became involved in the protection of nature around the issue of fracking and later expanded her work to help citizens of Toledo put forth the Lake Erie Bill of Rights. The bill asserted the lake's right to "exist, flourish and naturally evolve" and was meant to empower residents to bring lawsuits against polluters, such as big agrobusinesses whose use of chemical fertilizers has led to dangerous algae blooms on the lake, including the one in 2014 that caused the city's drinking water to become contaminated with toxins from cyanobacteria.[29] You cannot boil away microcystin, and it can cause liver damage and gastrointestinal problems.[30] Residents could not use their tap water for several days out of fears that it might cause illnesses.

But the Lake Erie Bill of Rights was overturned almost immediately; the Drewes Farm Partnership filed a lawsuit against it within twelve hours of the bill passing via city vote.[31] Not only was the bill overturned, but the Ohio legislature later added an amendment into the state code preemptively outlawing the ability of organizers to create other "rights of nature" bills.[32]

"What does that tell you?" O'Dell asks. "It tells you that they're very frightened of it and it must have power, because otherwise you don't pass preemption laws against something. And the other thing it tells you is they don't want you going and talking to your neighbors about a new idea. About a new way of doing things."

Having worked with CELDF for fifteen years, O'Dell has seen the limitations of using the legal system to enact change. It's given her pause and led the organization to try something new: hosting a Truth, Reckoning & Right Relationship event for the Great Lakes. In the first half of the event, over a dozen speakers offered their perspective on Lake Erie and how to have a better relationship with it. The second half was an art exhibition by Andrea Bowers hosted by the Museum of Contemporary Art Cleveland. Bowers created multiple paintings, sculptures, neon light displays, and a documentary for the show. O'Dell says the feedback from people participating in both events has been overwhelmingly positive. It's given her hope again.

"Does culture change the law, or does law change the culture?" It's a question she's been asking herself and other people. It doesn't mean CELDF is giving up on helping more communities draft rights of nature bills, but they're creating more programs that focus on connecting people in discussions about how we exist in relation to the environment around us. "If we could get enough people to understand we are nature, why would you need a law to say not to poison yourself?" she asks.

Neither O'Dell nor I think it's a coincidence that the rate of chronic illnesses has gone up in tandem with the quantity of pollutants affecting every aspect of our environment. It's so easy to be angry, to feel nothing *but* anger about these issues. To feel that it's a battle against not one but a hundred Goliaths.

But even in my anger, I don't hate any of the doctors, men and women, who have hurt or disbelieved me. I don't hate people who regularly eat beef, or have three cars for their family, or buy groceries with single-use plastic bags. And I don't hate the fact of floods, because water is just doing what it's always done.

No, the hottest embers of my anger are reserved for the byzantine systems permeating the healthcare landscape, the corporations that fail to take meaningful action on climate change, and the historic legacy of sexism, racism, and white supremacy that make the least privileged the most likely to suffer. I am furious at the health insurance companies that routinely deny care; private equity firms that gobble up hospitals, squeezing them for profit while depriving staff of resources and patients of adequate care; federal austerity measures that disrupt research agencies and cut funding for programs like USAID that provide HIV/AIDS support for people globally.[33] I rage against big agrobusinesses and chemical manufacturers that treat our waterways like dumping grounds; the history of redlining and Indigenous displacement that places racialized people in the closest contact with pollution;

gender inequality that means women suffer more from natural disasters and climate change than men.[34]

What to do with all this anger? It's not enough to cup the heat of it in my hands and stoke the fire with regular doses of injustice. I can't let it separate me from others, nursing grudges and planning vengeance. Healthy anger seeks connection because the ultimate goal must be change. Whether we seek to change the culture so that all people view themselves as part of nature, or to upend centuries of medical misogyny and racism, we can use our anger as fuel for the long fight.

I've occasionally been ashamed of my anger; who am I to be mad as an upper-middle-class white woman with more access to healthcare and greater financial security than many? How can I be mad about racism when I'm sure I've unwittingly perpetuated it, about sexism and ableism when those very ideologies are still at work within me? But then I think of Audre Lorde's speech, "The Uses of Anger: Women Responding to Racism." She called out the frequent displays of racism among white women, but she also said, "I am not free while any woman is unfree, even when her shackles are very different from my own."[35] Women's anger won their right to vote, sustained the Civil Rights Movement, catalyzed the fight for LGBTQ rights. Our anger sometimes causes friction between us, but as long as we don't let it turn to hatred and dehumanization, it can still help us move forward.

Nonviolent activist Barbara Deming also saw a space for anger, despite her pacifism. "Our task of course," she wrote, "is to transmute the anger that is affliction into the anger that is determination to bring about change. I think, in fact, that one could give that as a definition of revolution."[36]

That transmutation is what I'm seeking in sharing my story with others, here and in ordinary conversations. I don't want sympathy or pity or well-wishes. I want comrades who will lift a full-throated yell at this wretched status quo and help build something new.

PART FOUR

LANDSLIDES—GUTS—GRIEF

THE BREAKDOWN

Growing up on the southwestern shore of Lake Erie provided an abundance of water but a near-complete absence of hills. Hundreds of thousands of years ago, icy glaciers from the north repeatedly expanded and receded across the territory, compressing the earth and releasing water and sediment in their wake. In the last burst of glaciation, the Wisconsin glaciation, western Ohio was flattened. When those glaciers retreated, they left behind a river basin and Lake Erie, as well as a forested swamp that Indigenous groups of the region learned to navigate around.[1] These people also used fire to remove trees and grow grasses for food cultivation.

But when European settlers came to the region, the "Great Black Swamp" was a nuisance, taking days to traverse, with men and their horses "mid leg deep in mud."[2] More and more colonizers arrived and the wetlands became an ever-greater impediment. As environmental historian Dale Mize writes, "It was because of the forced relocation of Indigenous peoples that the progress made toward draining the Great Black Swamp roughly parallels the progress of resettlement by incoming Americans to the region."[3]

State and federal agencies worked to drain the swamp between 1870 and 1920. Once the water was removed, clay was processed into tiles and arable land converted for agriculture. The result was that by the time of my childhood, there were no federally recognized tribes present in Ohio, and only a few remaining patches of wetland.

But there were a handful of hills pushed into shape by human machines, such as the mound that rose above the forested wilderness of my backyard. Beeping, grumbling excavators crawled around a hill that, from my flatlander perspective, seemed huge, and they dug and dug. But the machines only excavated the center; sloping edges remained on all sides, scrubby weeds emerging from the clods of dirt. This giant hollowed-out hill proved to be one of the best places to play with neighborhood friends, especially as the patchy pine forest surrounding our home was progressively felled to make room for new houses. If we no longer had trees to climb, and farmers scared us off fallow fields, at least we had our mini-mountain. Running down its steep slopes, trying to hit targets with chunks of dirt or stones, holding races on the flattened center as if it were a track. It seemed to us an empty space, unused by anyone else. The result? We could make it ours without being told we were trespassing or damaging the dirt. Come winter, it would make the best hill for sledding.

We didn't know that it had been dug for a purpose beyond our amusement. The hole in the mountain was scheduled to be filled. Nearby marinas were dredging muck from the surrounding water, making deeper channels for boats. The slick mud and rock dug up from the shoreline wouldn't just be smelly with half-dead algae and hungry bacteria; it would likely contain pollutants like mercury and lead.[4]

My brother and I were banned from playing on the hill. It had never really been ours.

More than two decades later, thinking on the changes in my backyard landscape still brings as much sadness as nostalgia. The loss of the mountain was a small thing, but it came in the context of much larger hurts. My paternal grandpa, one of my favorite adults, had died of ALS just a few years before moving into that house. My first real pet, a guinea pig named Blanca, died of old age. My first cat followed shortly afterward, having been bitten by a venomous snake while outside. That cat was so damn mean to just about everyone, but I loved him. And to see even the landscape change—unmade by adults who didn't care if we kids needed wild, strange places to play—taught me how much loss is part of living.

Loneliness shadowed me all through middle school. Maybe it was the adolescent hormones, the changing friendships. But maybe, also, it was the collection of griefs I'd been accumulating. Sometimes the losses pile up so that you start to feel eroded, unsure of who you are without all those supports maintaining your identity. Then, that smallest new loss—a dirty hill, say—tips you over into real despair and it feels like the life you knew, the person you were, is simply gone.

It was a Saturday morning, a sunny day after weeks of rain in the small town of Oso, when the 190-meter-tall bluff collapsed.[5] The enormous mound of glacial till—sand and gravel above silt and clay—fell in two main chunks. The stages of the slide appeared on seismic instruments as sharp spikes within minutes of each other.[6] Eighteen million tons of material careened down the slope at forty miles per hour, collecting moisture from the North Fork Stillaguamish River and transforming into a liquid debris flow before engulfing the neighborhood of Steelhead Haven and destroying more than forty buildings in moments.[7] And because it was a Saturday, and only 10:37 a.m., many families were in those dwellings.

In the immediate aftermath of the Oso landslide, first responders from the local, state, and federal level assisted in rescue operations. But the sediments from the mudslide acted like quicksand, sucking down anyone who attempted to maneuver them by foot. The strategy shifted to using helicopters to spot survivors from above, then hoist them out.[8] Fourteen people were extracted from the debris field.[9] By Monday, officials were beginning to say that the work would shift from rescue to recovery operations; they were more likely to be finding bodies than pulling people alive from the mud.

Two landslide experts from the United States Geological Survey came to Oso to offer their expertise. One of them, Richard M. Iverson, had been studying the physics of debris flow for decades, including the aftermath of the Mount Saint Helens volcanic eruption that caused the largest debris avalanche in recorded history.[10] Even with such experience, he was disturbed by the scene.

"The landslide's distal deposit was a wood-laden chaos strewn with abundant evidence of obliterated human habitation. It looked like a debris field generated by a powerful tornado, except that much of it was encased in mud or quicksand that could swallow you up to your neck," Iverson wrote in a 2020 paper. "Hundreds of first responders were onsite, including EMTs, chainsaw operators, excavator operators, search dogs, and support crews who provided food, drink, and shelter for those working amid the horrific muck. . . . It was brutal work, which left even the search dogs traumatized because they detected the hints of human remains in many locations at once."[11]

That Monday, two days after the mudslide, I began to research it. Nearly three thousand miles away in New York City, I was working as an editor for Weather.com, the digital side of the Weather Channel. I'd already covered a half-dozen natural disasters. Floods, wildfires, hurricanes, and blizzards. But reading and writing about the Oso landslide shook me in a way the others hadn't. We expect weather to come from above, in all the forms that storms can take. I now had to reflect on the ways that weather profoundly influences land even after precipitation has ended. The air and the ground harmonize with each other, different chords from the same song: the surface composition of the planet affects the atmosphere and the atmosphere reshapes mountains, forests, deserts. When heavy rain raises groundwater pressure, which, in turn, triggers the earth to move, as it did in Oso, geology becomes a form of weather.[12]

All I could think about, covering that disaster from afar, was whether anyone saw the hillside collapse before it barreled into their home. Had they looked out the window at the beautiful river, enjoying the reprieve of sunshine after so much rain, and seen the wall of mud? Did they hear that initial roar of earth being dislodged? Was there time for fear before the force of the moving mud killed them? Or did they survive that immediate shock and suffocate under the dozens of feet of sediment? And one step further—how could the survivors go on living in the shadow of the sheered-off face of the hill? Ultimately forty-three people were declared dead, making the Oso landslide the second deadliest of such events in recorded US history, coming only after the 1985 Mameyes landslide in Puerto Rico.[13]

It was a strange job, being an editor for Weather.com. At least once a month there was natural disaster coverage to contribute to, which meant working early-morning or late-evening shifts to provide the latest updates from the National Weather Service and local meteorology stations. But my more regular work was creating photo slideshows and articles with clickbait headlines: "The Mysterious Sunken Ruins of Nan Madol" and "11 Spectacular Ice Caves Around the World" and "20 Astonishing Rock Formations Across the U.S." It gave me whiplash sometimes to go from making a photo slideshow of destination towns to finding photos of the victims of a natural disaster to write up a memorial.

The weather reporting at least felt important, even if it sometimes gave me nightmares. People needed to know about incoming storms and the ravages of climate change. I didn't think anyone was benefiting from most of my other work except for the corporation that owned the website. It was my first full-time journalism job, and it was nothing like I expected.

Or maybe it was only that my body—the body sitting at a desk in Midtown Manhattan each workday—was nothing like I expected. Even after getting treatment for my thyroid disease, after successful heart surgery, after a hormone-free IUD that I thought might cure my PMS symptoms, I still felt miserable. My stomach always hurt, food made me nauseous, and seemingly every week I came down with a bout of diarrhea. For years, my partner had been urging me to see a doctor about my stomach problems. He couldn't understand why I refused. The symptoms had been coming on so gradually for so long that I honestly believed nausea was a normal part of digestion.

And then I came across a celebrity diet book about gluten sensitivity and Celiac disease. I found it while volunteering at Housing Works Bookstore, a nonprofit providing support for people living with HIV/AIDS and combating homelessness. I initially thumbed through the book, then began reading closely as the author described her experience with thyroid disease and digestive symptoms. She'd lived for years with undiagnosed Celiac disease wreaking havoc on her guts. Her story bore an unpleasant resemblance to my own experience.

Gluten—proteins found in wheat, barley, and rye—lurks in the most unexpected places. Not only is it in bread and baked goods, but it can also be hiding in soy sauce and salad dressing and gravy and licorice. Putting the pieces together felt like acquiring cursed knowledge. Maybe I'd solved a puzzle, but what would it mean for how I lived?

I sat on our gray IKEA couch, still tainted with the smell of insecticide from the infestation we'd fought on first moving to this Brooklyn apartment. I turned over its back to face my partner in the kitchen. My throat felt tight, heart pounding with nerves at his reaction. "Hey," I said unsteadily. "I think—I think I need to stop eating gluten." And I started to cry as I explained my reasoning to him, even though he expressed relief that I was trying something. But I so very much did not want to modify my diet in such an extreme way. Me, who'd loved baking since age seven, who considered going to culinary school, who experimented with recipes for muffins on weekends. How many recipes had I learned from my mom, my babysitter, my aunts, and my grandmas? Brownies and honey cookies and pie crust and bread. Was I saying goodbye to ever baking with anyone again?

I felt awful at the thought of how much work it would be. But my partner would help, he said. At home, neither of us would eat gluten. He'd do everything he could to support me.

The best thing: within days it became obvious how much I was improving.

The worst thing: within days I realized this would likely be a permanent change.

It was such a small loss in the face of the disaster I'd been writing about for work. But the sudden change rippled outward to all corners of my life. Buying a new toaster, a new waffle iron, new cutting boards because we were paranoid about where crumbs might hide. I went from eating at Indian food carts around our office building to packing lunch or eating at the nearby salad place. And now there was the embarrassment of needing to *tell* everyone I was "gluten free," of asking detailed questions about how items in restaurants are prepared—if the French fries are cooked in shared oil that was previously used for gluten-containing mozzarella sticks or jalapeño poppers—of actually *returning dishes* to the kitchen because my salad came with croutons or there were rolls on the same plate as my chicken. I hated it, felt constantly uncomfortable, wished I could go back to being an avid eater of all cuisines without questioning every aspect of their preparation. Avoiding gluten wasn't such an impossible task; it was 2014 and the term was known. But some waiters doubted me, assuming I was following a fad diet. Others asked incessant questions about all that I could and couldn't eat. Meals are among the most essential communal experiences, and I no longer felt that I could easily partake.

In the span of two years, I had lost the image of myself as a "healthy" person, lost the innocence of never having had heart surgery, frequently felt as if I were losing my grasp on my own mind, between the depression and anxiety and sleeplessness my thyroid caused. Now I was losing a broad array of foods. My physical body felt better and better the longer I was away from gluten, but my soul continued to hurt. I didn't know to call it grief, then. I know now.

Social worker Courtney Wells was doing grief therapy with cystic fibrosis patients who were dying when she realized the role grief played in her own life. Diagnosed with juvenile rheumatoid arthritis as a baby, she'd spent a lifetime trying to cope with and make sense of her pain and limitations. She began presenting on grief at conferences and noticed people frequently cried early into the presentation, both with sadness and relief at recognizing something they'd felt but hadn't understood. This gap in understanding also proved important for healthcare providers.

"One of the doctors came up to us [after a presentation] . . . and he said, 'I've been misdiagnosing my patients with depression for forty years. And nobody has ever told me this,'" Wells says.[14]

We connected because she was doing further research on grief and autoimmune diseases (although she thinks that whatever insights she gleans will likely be applicable to chronic illness more broadly). Living with permanent disease is different from having a terminal condition and also different from losing a loved one. It's not a sudden, dramatic, permanent loss—at least, not always. Maybe it's being unable to go on vacation because one's symptoms are flaring, or realizing you can't pursue a certain job because it will be too demanding on your health, or losing certain foods, or losing friendships because people don't understand your new limitations. But then, sometimes it is a permanent, dramatic loss: a devastating diagnosis that means you can no longer walk unaided or live without pain. And there's also the ambiguous loss of knowing you'll likely be sick forever, but not knowing if/when it might get worse.

Wells created a presentation called "A Thousand Tiny Deaths" to illustrate the idea. "It often is the accumulation of these smaller things that add up over time," she says.[15] And if a person doesn't have the emotional space and support to cycle through those emotions—the denial, the anger, the fear, the sadness—the emotions can negatively affect our body. "Grief is supposed to happen, that's how our bodies and our brains deal with loss, and I think that is part of the disease process," says Wells. "But I think that we live in a culture, most of us, that doesn't allow us to do what is needed and to appropriately grieve."

Our talk reminds me of the way scientists sometimes describe the emotional component of climate change and mass extinction. Author and professor Elizabeth Rush attended one such talk with other faculty at Brown University. The audience listened to oceanographer Baylor Fox-Kemper talk about the deluge of water, and of information, coming with climate change. Near the end of his presentation, Fox-Kemper said, "We're going to lose some of the things we most value, and until we have the full emotional space around that, we're also going to have a hard time making sensible decisions."[16]

Writer Mary Annaïse Heglar echoes the sentiment. "People can't fix the problems if they don't confront them. We have to let people mourn so they can come out on the other side," she writes. "I never want to be the person who can look at so much suffering in this world and feel fine. It hurts because it's supposed to."[17]

How many people, upon first hearing of any of my illnesses, had suggested trying a specific diet, or supplement, or exercise that would surely cure me? How many had told me, "At least you don't look sick"? Even friends occasionally pointed out that I had a good life; I shouldn't focus on the health issues. Being confronted with the "forever" of chronic illness made them uncomfortable. Being honest about my grief was too heavy. There were so few opportunities to mourn what I'd lost to illness and what I feared I would lose in the future. It was, as it had been in middle school, so lonely.

However devastating they may be to humans in their path, landslides are a natural geological process. They play a crucial role in ecosystems, from nutrient cycling to creating conditions for certain species to thrive. They can occur in a variety of settings—essentially wherever a hill is too steep and material slides down it. But they're only abundant on about 4 percent of the Earth's surface during any given century: in mountainous, earthquake-prone regions, and landscapes that have been heavily modified by humans for things like roads and mines and irrigation for agriculture.[18] Rain is the main climatic trigger for landslides, as was the case in the Oso landslide.

But the forces that push a stable slope into an unbalanced one are myriad and go far beyond precipitation. The ground itself is an active player. Rocks can snap as a result of erosion or the growth of tree roots; conversely, the roots of plants sometimes anchor the soil. As for all the creatures that live and die within the ground, be it soil or rock—that's even more complicated. Multiple researchers have called soil a "black box" to convey how little we have discovered of its microbial diversity.[19] And because we don't know what lifeforms are there, we can't know what role they might play in cementing a peak in place.

What we do know is that heavy precipitation events are increasing in the US at the same time that humans are transforming ever more of the landscape.[20] We've modified nearly 95 percent of the Earth's landscapes and are still developing new roads and mines and logging projects and houses.[21] All this terraforming pushes hills to the edge of their tolerance; with every new stressor, the slopes inch closer to a hazardous slide. And fatal landslides triggered by construction, illegal mining, and hill cutting are increasing. Between January 2004 and December 2016, 55,997 people died in landslides unrelated to seismic activity.[22]

Even agriculture is contributing to landslide occurrence, says geologist David Petley, which was a surprise to me, as someone who comes from a town where corn and soybean fields are flat as paved roads. But there are many places where hillside agriculture is a necessity: the Alps, the Andes, the Kyushu Mountains of Japan.

"Traditional agriculture was pretty adapted for it," Petley says. "If you go into the Himalayas, for example, people grow rice on astonishingly steep slopes, but they've done that by terracing them."[23] Problems occur when farmers neglect those carefully maintained terraces. Then the water management breaks down, and the slopes collapse. In other cases, small farms turn into much larger corporate farms and the fastidious monitoring of soil conditions disappears. "[They're] using farming techniques which generate very high yields very quickly, but with techniques that are not good for preserving slopes, so quite often you see landslides occurring in those environments," Petley says.

But landslides don't have to be deadly, even in densely populated areas. Hong Kong has a sophisticated system of rain gauges, precipitation forecasts, and landslide alerts, put into place after the region experienced two notable fatal landslides and many smaller ones.[24] Petley tells me about his work with the Hong Kong government to create this system and how it's now preventing deaths. He also describes a near miss in Taiwan when a typhoon generated multiple landslides, which nearly overran him and a colleague. These days he tends to monitor things from afar, with a long-running blog about global landslides. He and his colleagues are also researching the human factors that contribute to landslide vulnerability. Our conversation leaves me thinking about how quickly something that appears immovable can suddenly break down, and the investment required to understand this breakdown and either prevent it or mitigate its effects.

Petley says that even given what we know about the physics and geology of slides, significant questions remain unanswered in the field. The connection between where an earthquake hits and where landslides will be triggered is still being investigated, as are the dynamics of massive slides, the kind that could happen if entire mountainsides collapsed. The trouble is that we don't know when or where these catastrophic landslides will happen, but even if we did, finding instruments that could survive the immense forces to study them is a real conundrum.

I have only ever visited mountains, not lived among them. When I hiked Mount Tongariro in Aotearoa/New Zealand, and Haleakalā in Maui, I was astonished by the variation in terrain across a mere dozen miles. In the former, ice-covered alpine peaks offered views of vibrant geothermal springs before a descent in elevation brought me to scrubby grasslands and a forest. In the latter, what started as the Martian landscape of a volcanic crater turned into lush greenery from mist blowing over the collapsed edge of the crater. I left both mountains with bruised toes and blisters, as well as an appreciation for the incredible abundance of habitats. I understand why humans choose to live in and near mountains, despite the risks of landslide and avalanche. It's comforting to believe something so solid will last forever.

But nothing on Earth is permanent. Even rocks go through cycles. If human behavior obviously affects the water, the air, and the atmosphere, why was it a surprise to me that we're changing the ground as well? Perhaps because it is easy to overlook the incremental buildup of destabilizing forces. Human life is so short on the geologic timescale that it's almost impossible to fully comprehend the transformation of rocks. Or maybe something in human nature—or at least my nature—rebels against the idea of impermanence in a landscape as steady as a mountain. The weather and climate and wildlife around them may change, but surely those ridges do not.

They do, though. Of course they do. As science fiction author Octavia Butler put it, "The only lasting truth is Change."[25] The hillside collapsed on Oso. Many hillsides inundated northern Venezuela and killed tens of thousands after heavy rain in 1999.[26] Landslides that followed earthquakes have collectively killed thousands in China and Indonesia and Tajikistan. Let's let go of the illusion that the earth under our feet and above our heads never moves.

In reconceptualizing my relationship to mountains, I find it useful to think of the ways that these ranges, seen from above, look surprisingly similar to the hills and valleys within our guts: those torquing, changeable layers of the digestive tract. The wrinkled ridges, the density of surface area, the way they create their own ecosystems. Mountains exist on different scales than our GI system, both spatial and temporal, but they have moments of sudden change, too. I think of how our insides are so complicated we're still trying to understand which microbial communities inhabit the mouth versus the stomach versus the large intestine, and then I think of all those unanswered questions about landslides.[27]

How many landslides are caused by poorly planned human activity, or the unforeseen consequences of fossil fuel consumption? How many digestive diseases are the result of consuming pesticides along with grains, or having a diet that lacks fiber, or inhaling smog, or experiencing the stress of losing one's job and source of income?

What I want to know (but what no one can tell me) is whether my thyroid disease predisposed me to gluten intolerance, or if something wrecked my gut long ago, and my thyroid was only the first signal that something had gone wrong. I want an answer because it's hard to live with all the unknowns, this ambiguous loss, the endless wondering if I could've done something different and if more will go wrong in the future. I ask the same questions of our CO_2-laden atmosphere and human-mediated habitat destruction: How many more species will die? When will the disaster come for my home? Who among my circle will die young, and can I still prevent that?

Here is my magical thinking at work: maybe if I fully understand the processes, I can finally accept all that's happening. I'll be done grieving. But grief, like chronic illness, is not a condition one recovers from. It's something you learn to live alongside.

The first mistake was going gluten-free before talking to a doctor. The second was deciding I really did want a doctor's feedback.

I chose the gastroenterologist because she was a woman. I'd had enough of men, the endocrinologists who didn't believe me when I related my symptoms, who berated me for asking too many questions and questioning their judgment. I forgot that a woman misdiagnosed my heart arrhythmia, a woman brutalized my uterus when inserting my IUD. I didn't want to accept the obvious lesson: in the doctor-patient dynamic, it's rare to have any power at all as a patient, no matter the gender of the doctor.

Things didn't start well. I recounted my symptoms, the way a gluten-free diet eliminated all of them. I told her I thought it was Celiac disease. She listened, but said she couldn't test for Celiac disease unless I was still eating gluten. She wanted to do an endoscopy, in which a camera would be threaded down my throat to peer into my stomach and the first part of my small intestine, where damage from Celiac disease appears. But in order for the test to be accurate, I'd have to go back to eating gluten. This is called a "gluten challenge."

It had only been a month since I excised barley, wheat, and rye from my diet, but the thought of bringing it back into my daily meals felt like becoming an arsenic eater, trying to reap benefits from the thing that may kill me.[28] Maybe if I treated it as a fond farewell to my favorite foods, it wouldn't be so bad. A baguette from a French bakery, slices of pie, sandwiches on bread that doesn't crumble to dust with the slightest pressure.

But what I'd hoped would offer closure only made me miserable. I was constantly nauseous, running to the bathroom a dozen times a day to dry heave or have another bout of diarrhea. The symptoms crept beyond my digestive tract: exhaustion, allergies, a brain muddied by distraction, an inability to lift my spirits. It was so unbearable that I called the doctor and asked if I could end the challenge early. It had been nine days. She said that was good enough.

The endoscopy happened at the office, in a garden-level room on a mustard-yellow reclining table. I laid on my side, a plastic guard holding my tongue down and teeth apart. Then sleep, followed by a hazy awakening, followed by an afternoon of my stomach bloated by air I didn't create. The scope took a sample each from my stomach and small intestine, tiny fragments of lining sent out for biopsy.

When I returned to the office a week later, the doctor was pleased to say I didn't have Celiac disease and should therefore eat gluten from time to time.

"But it makes me feel terrible—even a little bit, like soy sauce in a stir fry."

"You might have IBS," she said, dismissive. She was more interested in the fact that I had unusually elevated liver enzymes. Maybe my stomach symptoms were a red herring.

I left the office of this female doctor, uncertain as to whether I was relieved or annoyed. No matter what she said, I resolved to go fully gluten-free.

Over the coming weeks, my liver returned to normal. A red, itchy rash that had plagued my butt for ages disappeared.

Years later, in a new city, with a new gastroenterologist, the doctor asserted that I most likely had Celiac disease all along and confirmed I must avoid gluten. Liver enzymes are often slightly elevated with untreated Celiac disease; the rash I had may have been dermatitis herpetiformis, a condition triggered by gluten in people with Celiac disease.[29] To cap it off, a gluten challenge must be at least two weeks to produce valid results; I barely made it past the one-week mark.

There was little grief in an official diagnosis; I had long ago learned how to avoid risky foods. But sadness did arise at the thought of my confusion on that first doctor's couch, my sense of betrayal when she told me to keep eating wheat. How did she miss so many signs? Why didn't she listen to what I told her? It's exhausting to maintain such vigilance, to do the work of questioning doctors. Because even if she didn't understand the severity of my symptoms, I do. I have to live with them.

The worst gluten exposures, while infrequent enough to damage my small intestine, have left psychic scars. A friend cooked dinner for me and afterward I was swallowing bile on the metro. A kitchen drawer in a new apartment contained the previous owner's crumbs, which sprinkled our cutting board and got in the food. Days of liquid guts, unable to leave the apartment. Worst of all was during a vacation in New Zealand, when a restaurant served me the wrong dish. I was so sick later that we were trapped in a café for hours; every time I took even a sip of water I was forced to run to the bathroom. When that café closed, we went to another, till it seemed that every scrap of nourishment and moisture in my system was gone. Only then, some eight hours later, did we drive on to our next destination.

This is the extent of disruption that accompanies my accidental gluten exposures, and the symptoms last for days. The world of safe food is permanently shrunken for me.

Doctors expect unquestioning compliance with their judgment. And I understand the rationale: most patients do not have years of medical study and licensing credentials at their disposal. With my current education level, I could no more conduct an esophagogastroduodenoscopy than I could design an architecturally sound strip mall. What I do have is an intimate knowledge of my own body. Never would I choose to remove whole categories of food from my diet without some strong motivation. Nor would I agree to invasive testing if I knew the results would be inaccurate without some other conditions being met.

I think what many doctors do not understand when they see patients with chronic and complex illnesses is how much grief we already hold in our bodies as the result of our diseases. The last thing we need is for their oblivious or dismissive behavior to inflict yet another wound. They can't know what might serve as a tipping point to push us into despair. Sometimes the patient doesn't know either.

MICROBIAL MAYHEM

Mysteries abound in the appearance of life on Earth. The age of the oldest fossilized evidence is debated, with dates ranging from 3.5 billion to 3.7 billion years.[1] The fossils in question are microbial mats, which formed rocky structures called stromatolites. To see them today, you might just think they were rocks, domed and bubbled like half-melted marshmallows. But they were actually a complex ecosystem of bacteria, archaea, protists, viruses, and sediment.[2] The layered fossil stromatolites are still visible today in South Africa and elsewhere. Joan Bernhard, a geobiologist at Woods Hole Oceanographic Institution, notes that the stromatolites "were one of the earliest examples of the intimate connection between biology—living things—and geology—the structure of the Earth itself."[3]

That relationship has continued across Earth's history, taking a number of surprising forms. Researchers have found that the burrows of ancient sea worms called polychaetes affected the chemical composition of mudstones on the seafloor.[4] Other shelled sea creatures, like foraminifera, are a key component of the sedimentary rock limestone. And researchers have recently speculated that microorganisms may have a crucial role to play in the movement of rock that leads to landslides.

In 2009, southwestern China experienced a disastrous rockslide that killed seventy-four people.[5] The Jiweishan slide had no obvious trigger, like heavy rain or earthquakes, though the mountain had been mined for coal and hematite in the past, which caused some instability in the overhanging rock. Researchers hypothesized that acid rain—which has a pH around 4.2 to 4.4, compared with normal rain's pH of 5.6—slowly turned a bed of black shale into a weak and slippery talc.[6] The rainwater likely contained abnormally high levels of sulfuric acid and nitric acid due to pollution released by coal burning, one of the primary sources of power in the region. Sure enough, a lab experiment showed that this transition from one rock to another was chemically possible.

But that might not have been the only, or even the main, factor in the rock collapsing. The same researchers suggested that microorganisms like bacteria and fungi may also have contributed.[7] Carbon and phosphate present in the shale were already feeding the hungry microbes when the acid rain brought an influx of nutrients like potassium, sulfur, and nitrogen, as well as oxygen. The growth this prompted may have allowed the microbes to more rapidly decompose the rock and weaken it to the point of collapse.

Another group of researchers applied an even more detailed approach to the microbial life of soils within and outside landslide zones in Italy.[8] By analyzing the DNA of bacteria in landslide and non-landslide soils, they found substantial differences in the microbial networks of each type. They concluded that the species living in the landslide zones could well have contributed to the slides, though much more research will need to be done to understand how this happens. But they emphasize that focusing only on macroscopic factors of landslides ignores a key element of the way rocks and earth break down.

The question I'm left with is how large-scale human activity affects these microbial communities responsible for degrading or solidifying sediment. Even beyond the movement of plate tectonics—the subduction and volcanism and uplift—stone and soil are not inert. Life thrives between the cracks, to such an extent that we don't have names for most of those organisms.[9] And if we don't even know what's there, how are we to infer what air pollution, and agriculture, and chemicals, and global warming are doing to them?

The peak era of stromatolite distribution was the Proterozoic, 2.5 billion to 541 million years ago.[10] The microbial mats thrived in waters across the world, sometimes even surviving when animals arrived and started grazing, burrowing, and disrupting the shallow seas.[11] But changes in water chemistry and fluctuations in sea level (two components of climate change that are increasingly being observed) played important roles with regard to when these organisms excelled and when they died off.

Even after all this time, some stromatolites have clung to life in extreme environments, like the hypersaline Hamelin Pool in Shark Bay, Australia. Shark Bay is a UNESCO World Heritage Site and one of the world's most diverse existing stromatolite systems; among the abundance of life, from dugongs to bottlenose dolphins, the stromatolites are considered the most recognizable feature. But for all their hardiness, stromatolites are under threat from climate change. Increased storms and heat waves and ocean acidification are changing the conditions that hitherto have allowed the microbial mats to endure. Researchers still don't understand the impact of such ecological upheaval on the stromatolites, but they say we can't take their resilience for granted. We have plenty of fossil evidence from the past that these strange organisms can and do go extinct.

Most of the time when bacteria face new conditions, they respond with incredible adaptations.[12] There's a reason this domain of life has never disappeared from the planet since their arrival; bacteria proliferated during mass extinctions that killed off animals.[13] Unfortunately, their survival strategies can have negative impacts on human health. Scientists recently proposed a new field of study called "disaster microbiology" to investigate how fungi and bacteria affect humans in the aftermath of tornadoes, hurricanes, droughts, and other extreme events.[14] Microorganisms will adapt to the changing planet, whatever damage we inflict; those adaptations are already coming up against our own ability to survive.[15] Just one example is the waterborne *Vibrio vulnificus* bacteria that can cause lethal vibriosis infections in humans exposed to the bacteria through open wounds or eating raw shellfish. The bacteria's range is expanding thanks to warming oceans and intense hurricanes.[16]

Humans and bacteria coexist on a spectrum from the acutely harmful *Enterobacteriaceae* (which includes *Salmonella* and *E. coli*) to the endlessly helpful *Lactobacillus* that help us digest food and make cheese and wine. One bacterium in particular might be the ne plus ultra; without it, the human form we've taken wouldn't exist.

The earliest known eukaryotes (organisms that contain a nucleus) date back 1.45 billion years, and back then they already had mitochondria—an organelle you might remember as the "powerhouse of the cell."[17] What's truly mysterious about mitochondria is that they were free-living bacteria doing their own thing before they became part of their host. And what exactly was the host that absorbed the mitochondria, creating its endosymbiotic relationship? We still don't know.[18] We don't know how or why or even when these two organisms came together—but mitochondria have been crucial to eukaryotic development. This microbial partnership evolved only once that we know of, and most everything we think of as "living" shares a certain number of genes in its mitochondrial DNA. Nearly all our human cells are part bacteria.

The relationship doesn't end there. Our bodies teem with unseen life, to the point that bacteria cells outnumber the cells of the human body by an order of magnitude.[19] Microbes affect how we digest food and absorb calories, whether we fall prey to skin infections, and how our immune system is trained to detect pathogens.[20] The myriad ways our fate is written by our microbiomes is only just beginning to be understood, but there's no doubt we exist in a delicate balance with an unseen microscopic world.

Newborns are exposed to this plethora of fungi and bacteria from the moment they leave the womb. But the type of bacteria that colonize their skin and mouth and gut depends on how they enter the world and what they first eat.[21] Vaginal birth provides different bacteria than a Cesarian delivery; babies fed mostly breast milk but supplemented with formula develop different gut bacteria from babies fed exclusively formula.[22] And pregnant people vaccinated against the flu, pneumonia, and meningitis can pass on those antibodies through breastfeeding, providing the newborns with a greater immediate defense against illness.

I was my parents' first child, delivered via C-section after a difficult labor, and was fed on breast milk and formula. Without surgical intervention, my mom and I both would've likely died in childbirth. Without formula, my mom couldn't have gone back to work at the end of her maternity leave.

These two interventions—surgical delivery and infant formula—have undoubtedly saved lives. Research on neonatal and maternal mortality in 194 countries found that when the rate of C-sections rises to 19 percent of all births, mortality rates decline.[23] The story of infant formula is more complicated, with some research showing that the introduction of Nestlé formula in low- and middle-income countries in the 1970s and '80s came with a 27 percent increase in infant deaths, due to families mixing the formula with unclean water.[24] In places where the water was safe enough for use in formula, the mixture provided nutrients essential for growth and development and allowed parents to more easily go back to work. But growth is not the same as long-term thriving. Breast milk reduces the likelihood of allergies and lowers the risk of leukemia, along with providing the other benefits I've already listed.[25] Even after decades of tweaking to make formula ever more nutritious, it still can't match the complexity of breast milk. Exclusively breastfeeding for the first six months of a newborn's life is recommended by the World Health Organization in almost all cases (though personal and economic circumstances may make that impossible).[26]

From landslides to bacterial infection to infant health, so much is linked to food: what each organism eats, how the nutrients are produced, how much is at everyone's disposal. We are locked in a sometimes antagonistic, sometimes beneficial partnership with microbes, and they have a similar relationship with inorganic materials of the earth. The ways humans have responded to the challenges of food production, disease, and environmental change are the pillars of our societies. And for virtually all human history, lack of food has been a critical issue for survival. Famines even occurred throughout the nineteenth century, including in Finland, India, and Ireland.[27]

Scholars say the Industrial Revolution was an inflection point that went hand in hand with improvements in agriculture.[28] The Industrial Revolution allowed fewer farm workers to produce greater amounts of food, thanks to the increased use of chemical fertilizers and purchased feed for livestock.[29] These strategies increased productivity without an equal increase in land use for cultivation and animal husbandry. There was also a decrease in freight rates, so food could be shipped across oceans much more cheaply. All these disparate elements contributed to the astonishingly steady growth of agricultural production from the 1870s up through the 1930s, with the only dip coming around the First World War and possibly the Great Depression. Historians suggest that agricultural production was essentially constant before the Industrial Revolution.[30] The growth trajectory afterward is the true anomaly.

The end of World War II brought an increase in the use of tractors and other machinery, as well as the "Green Revolution." This suite of public and private research initiatives led to the creation of high-yield, diseases-resistant varieties of cereal grains—seeds that transformed the global agricultural landscape.[31] In India, for example, the implementation of these highly productive crops led to a tripled production of cereal grains with only a 30 percent increase in farmland.[32] The world we've inherited in the twenty-first century is one brimming with so much food that almost a third is never consumed, partly because distribution is far from equitable.[33] Famine has become less a consequence of failed agricultural practices and more the result of warfare and genocide, such as in Gaza and Sudan. But it may not be long before climate change also transforms food availability: the UN's World Food Programme estimates the risk of hunger and malnutrition could rise 20 percent by 2050 due to the effects of climate change.[34]

But food waste, unequal access, and rising temperatures and CO2 levels are far from the only problems with our current food supply. The use of chemical fertilizer and pesticide, combined with concentrated animal feeding operations, leads to devastating unintended consequences. Pesticides increase the amount of heavy metals in soil, including lead and cadmium.[35] Farm workers exposed to those pesticides without proper protective gear face endocrine diseases, neurologic disease, reproductive issues, and cancer.[36] Their exposure also transforms their gut microbiome, with possibly deleterious effects.[37]

It's not only those in closest contact with pesticides being affected; the chemicals are released into the environment through the air, through runoff, and through groundwater leaching. They've been found in marine and terrestrial animals.[38] Other problems occur with the overuse of chemical fertilizer: runoff creates algal blooms in nearby bodies of water, which cause their own health problems—like the poisoning of the water in Toledo, Ohio, due to dangerous blooms in Lake Erie.[39]

Tvisha Martin was in high school when officials warned against using Toledo's water in 2014, but by the time she heard about it, she was already sick.[40] The experience led her to study water quality as an undergraduate, measuring levels of *E. coli* in Maumee Bay, the westernmost edge of Lake Erie. Then she had a random conversation about soil health, and she realized that to understand the source of the water contamination she'd been monitoring she'd have to learn more about agriculture and soil. That's when she decided to apply for a graduate program at the Soil Health and Ecosystem Ecology Lab, led by Christine Sprunger.

According to Sprunger, an associate professor of soil health at Michigan State University, soil management has historically been about fertilizer use and the measurement of nutrients like phosphorus and nitrogen. "But in the past 10 to 15 years, there's really been this new recognition that it's the living aspects of soil that are so critical for not only food production, but also providing other really important ecosystem services," Sprunger says.[41] Healthy soil can store carbon, prevent floods during heavy rain events, and even produce more nutritionally dense food. But the researchers who look at the living side of soil face so many mysteries: genetic sequencing test results regularly come back with unknown microorganisms. Faced with such an immense multitude of bacteria and fungi and the daunting challenge of identifying all of them, Martin and Sprunger decided to instead look at *slightly* larger lifeforms: nematodes, also known as roundworms.

These microscopic wigglers all sort of look the same, unless you've had the specialized training the two women received in the Netherlands. There, they learned to identify different species based on their mouthparts, and to place them in different categories based on their diet: the bacterivores, the fungivores, the omnivores, and the carnivorous predators. Some nematodes are parasitic to plants, like the soybean cyst nematode, which, as the name suggests, damage soybean crops. Other parasitic nematodes cause GI infections in humans, traveling from the soil to the skin to the digestive tract. A report from 2006 suggested that about half the world's population is infected with nematodes, causing staggering levels of illness and death.[42]

In healthy agricultural ecosystems, there is a variety of nematodes, with the top-level predators balancing out bacterivores and fungivores. But when there start to be too many bacterivores, it means underground decomposition is happening too quickly, and the chemicals in the soil are being released straight into the atmosphere without much carbon sequestration. These imbalances in the nematode population happen in conventional agricultural fields: tilling chops up the bigger predatory nematodes; pesticides alter their diversity; monocultures cause a disturbed food web in the soil. And what happens to the nematodes is reflected in the other ways the soil suffers. It becomes more vulnerable to plant parasites, less able to hold water in heavy rain, and more likely to suffer during droughts.[43] If farmers switch to regenerative practices such as no-till, they can undo these harmful effects in the soil and produce higher yields within three to five years.[44]

"We're already experiencing climate change, and it continues to just get worse with time," Sprunger says. "So it's really critical that we shift our agricultural systems as rapidly as we can."

The trouble, as she points out, is that some of the effects of climate change are here. More intense rainfall, more drought, more unpredictable weather patterns. These changing conditions can be advantageous to certain bacteria, such as *Salmonella enterica*, which requires humidity for growth. Scientists already knew that the combination of summer weather and irrigation allows *Salmonella* to propagate, and new research has found that high humidity allows *Salmonella* to more easily infect leaf lettuce.[45] With some forty-eight million cases of foodborne disease recorded in the United States each year, and *Salmonella* increasingly being transmitted by fresh produce, it's likely that we'll continue to deal with health impacts of our food production system as climate change progresses.[46]

The World Health Organization estimates that 420,000 people die of food poisoning every year, and death is hardly the only health impact.[47] Even if the encounter doesn't kill you, it can cause lingering health effects, especially if you're very young or very old, or if your immune system is compromised in some way. *Campylobacter* infections from raw and undercooked poultry can lead to Guillain-Barré, a rare condition in which the immune system attacks the body's nerves, causing symptoms like numbness and paralysis.[48] Bacterial infections like *Salmonella* and *E. coli* can trigger inflammatory bowel diseases like Crohn's and ulcerative colitis.[49] Raw milk can transmit tuberculosis, Rift Valley fever, and a number of other diseases. Foodborne illness can even trigger autoimmune thyroid disease. And given the challenges in tracking all but the most severe outbreaks of food poisoning even before the Donald Trump administration began dismantling the FDA, people may never realize that the bout of stomach illness they experienced led to long-term disability.[50] But as feminist scholar Alison Kafer writes, "We often cannot cleanly separate *being* disabled from *becoming* disabled, from the literal causes of disablement or debilitation. And these causes are often traumatic sites of violence, both individual and structural, both singular and chronic."[51]

It was the first year I felt confident in my ability to travel regularly for freelance reporting. I didn't know that it was also likely to be the last.

I'd learned long ago that traveling with my own food was the best strategy to ensure I didn't go without eating. Cups of gluten-free oatmeal, freeze-dried camping meals, protein bars, dried fruit. Half my luggage might be taken up by food, even on fellowships where meals were provided. When possible, I would book places to stay with partial kitchens so I could cook.

This was the case for my summer visit to British Columbia, where I was going out to an active archaeological dig in the Broken Group Islands, part of the territory of the Tseshaht First Nation. My gear included a tent and sleeping bag for the dig, but I was in an Airbnb the preceding night. I took advantage of a small farm stand to get fresh vegetables for dinner.

Did I not wash them enough? Not cook them thoroughly? Not wash my hands? I can't tell how the contaminant crept into my meal, only that one did. I ate, cleaned, prepared for my morning departure, headed to bed, only to be stricken with terrible stomach cramps, nausea, and urgent diarrhea.

This continued from late night till early morning, not as severe as a bad gluten exposure but still plenty miserable. I decided to avoid most food for the day, not that I had much appetite. I got through the drive, the stomach cramps, the boat rides the next day. It turned into one of the most incredible experiences I've had in my life, the food poisoning a minor blip of discomfort.

But that wasn't the only food poisoning that happened in 2019. Next came the illness in Indianapolis, after I ate at P. F. Chang's.

This had been one of my safe chain restaurants, for the care they take with gluten-free dishes. I'd had a productive day of interviews and was excited for a conference a bit further south in Indiana the next morning. I ate my meal and brought leftovers back to the Airbnb, thinking they'd make a decent lunch the next day.

But once again, the stomach cramps were followed by nausea, both severe enough to keep me awake most of the night. In the morning, I called my sources to say I was sick and to apologize for missing their conference. It was another exhausting drive, but this time to my own apartment, my own bed. I'd had wrist surgery a month earlier and still wore a brace, so I took the additional comforts of my space whenever I could, even if it meant missing out on a reporting opportunity. I assumed this would be like the last time—a mild case that disrupted my digestion for a few days but would be gone before the week's end.

Only I didn't improve. I got worse. When my in-laws came to visit a week later, I could only eat a few bites of soup at our favorite restaurant. The arrival of Thanksgiving came with delicious dishes, but the stuffing and green beans and pie all made me feel terrible. I began subsisting on rice and scrambled eggs and sweet potatoes, forcing myself to consume calories even when food was physically painful to consume. Nausea plagued me. The diarrhea rarely stopped. I couldn't seem to find a single food that sat well in my stomach. I lost weight, a lot of weight. Eight pounds, ten, fifteen, till none of my pants fit and I felt skeletal. Worse still, something about the rapid weight loss and GI symptoms seemed to be triggering issues elsewhere in my body. My joints ached like I had the flu. Low-grade fevers came and went. My left eye turned bright red, not from an infection but from inflammation. An eye doctor diagnosed uveitis and prescribed steroid drops; he urged me to be tested for inflammatory bowel disease, because if I were sick with Crohn's or ulcerative colitis, my uveitis would likely keep coming back till I had a more global treatment.

My primary care doctor tested me for parasites, fungal infections, bacterial infections, Lyme disease. It all came back negative.

When I finally got in to see a GI doctor, a little more than two months since my symptoms began, he made room in his schedule for an urgent colonoscopy and endoscopy. I didn't even mind having to do the bowel prep, a day without food and the purging caused by the salty mixture. I had no appetite and near constant diarrhea anyway. This really wasn't much worse.

When I was wheeled into the surgical room, I was prepared to be told I had some terrible disease. What I hadn't prepared for, in the groggy aftermath of awakening, was that the doctor hadn't seen anything obviously wrong. He'd taken biopsies and would check for microscopic colitis, but he genuinely didn't know why I was so sick. I lay in the hospital gown, still hooked up to an IV, trying not to cry. All because a doctor told me I *didn't* have some awful, incurable disease. Because it meant that if he didn't know what was wrong, there might be no way of treating me.

Illness, when it comes without a name, feels like being trapped in a murder mystery. Something dramatic has happened: the death of the former body. The work is to see the players, accuse, solve the crime. So the professionals pull out diagnostic tests and scans; if we're lucky, insurance companies don't get in the way, and an answer appears, and we're awarded treatment. But that's not how most crime investigations play out in the real world; nearly half of murders go unsolved in the US.[52] And often, doctors don't manage to diagnose someone's illness.[53] We learn to live alongside the corpse, nudging it every now and then, as if it might still offer up the killer. We can't escape the grief of loss, because the loss might never be resolved.

ALL WE HAVE TO LOSE (OR SAVE)

Imagine a small village in the mountains of Nepal. The residents live off subsistence farming and struggle to bring their products to a larger marketplace. Access to healthcare and education is nearly impossible. Then one day, the government cuts a road through the mountain, not far from the village. The road construction gives people jobs and money. More villagers begin moving closer to that road, erecting houses and little shops right alongside. Suddenly there is a way to receive quicker medical assistance when it's needed; the road brings economic prosperity; it makes life easier. But the road also destabilizes the hillside, and Nepal exists in a monsoon climate with annual heavy rains. Landslides begin killing more people. This is the price of living alongside the road.

This, as geologist Dave Petley explained to me, is an example of the complicated facets of vulnerability. "For the local people, increasing your risk of dying in a landslide while reducing all those other risks is a trade-off that is probably well worth making," Petley says. "Lots of people die in landslides. Even more people die because of a lack of access to healthcare or in childbirth or essentially because they can't feed their family."[1]

Here's another scenario: members of the Hmong ethnic group flee China to avoid forced assimilation. They end up in Laos and Vietnam and other parts of Southeast Asia. Next, they're caught up in the American political and military campaigns in Vietnam and Laos, and must flee to Thailand, spending as long as two decades in refugee camps.[2] During and after this period, they also come to the United States as refugees, with more than sixty thousand ending up in Minnesota.[3] They have access to welfare support and public housing, as well as to other Hmong in the community that has now taken root. And they're far from the only refugees and immigrants from Southeast Asia to come to Minnesota: Cambodians, the Khmer, and the Karen people have all moved to join communities in the state. But new research has found that the dietary shifts that come with resettlement can cause significant changes to the gut microbiome.[4] Within just nine months of arriving to the US and consuming a Western diet higher in sugar, fat, and protein, the Karen immigrants start experiencing profound shifts in their gut microbiome. And the longer people remain in the US, the less diverse the gut microbiome becomes, as the research has shown in multigenerational Hmong families. Certain strains of bacteria responsible for digesting dietary fibers are lost. Not only are these changes compounded the longer a family stays in the United States, they may also contribute to higher levels of chronic disease. But in the face of displacement and persecution elsewhere, perhaps this, too, is a risk worth taking.

Individuals make sophisticated decisions about when and how they undertake certain risks. But choice can also be constrained by broader circumstances: in the same geographic location, vulnerability may vary between different communities, and it can change over time. When I understood the risk posed by eating gluten, it was a relatively easy choice to remove foods containing it from my diet. Yes, it can be difficult in terms of socializing, and there is undoubtedly a financial impact: gluten-free products are much more expensive than their gluten-containing counterparts. But I have the means to afford those foods and the ability to seek them out. The same cannot be said for all people.

The ethnic group with the highest known rate of Celiac disease, at 5.6 percent of the population, is the Saharawi, a historically nomadic people who settled around Western Sahara—now a contested territory held by Morocco.[5] Some 173,600 Saharawi people live in refugee camps in Algeria, where they've been for fifty years.[6] Around 88 percent of the population is either at risk of food insecurity or already food insecure—which means that even if a child is diagnosed with Celiac disease, wheat bread might be one of the few sources of food. The result is children suffering from chronic diarrhea, stunted growth, depressed mood, and iron-deficiency anemia.[7]

Decisions regarding who we allow to remain vulnerable and who is offered resources to avoid some of that risk say as much about our social priorities as they do about the geography and climate of any given place. It is possible to provide nutritious food that meets the dietary restrictions of sick children in refugee camps. It is possible to choose the health of the planet over the profit of mega-corporations. People can demand that their government invest in climate resilience and renewable energy rather than expanding military infrastructure. But humans are complex, and we don't always (or often) agree on the best methods for taking these actions. We live in a culture that values some people's needs and comfort over others, that values wealth accumulation over the equitable sharing of resources. And, as I learned throughout my twenties, we live in a culture that sees the sick and disabled both as less worthy citizens and as an infinitely exploitable group.

I found the functional medicine practitioner online, seduced by patient reviews proclaiming "life-changing" impacts. While functional medicine purports to treat the patient holistically and get to the "root cause" of disease, practitioners sometimes eschew well-vetted pharmaceutical medications in favor of selling their patients expensive supplements. Logic told me to expect little of someone who charged exorbitant fees and whose treatments wouldn't be covered by any insurance plans, but desperation overrode all other considerations. I'd tried anti-emetics, anti-diarrheals, antacids, a huge pill to reduce stomach inflammation. Nothing helped for long. It felt as if food, so necessary for sustaining life, were killing me. Sometimes I laid awake at night, exhausted but unable to sleep, and charted the self-replicating hydra engulfing my torso: the painful muscular spasms migrating from the center outward; fire in the lower left corner; a pickax near my spleen; the constant, relentless nausea. Maybe this guy would find something the other doctors didn't. If he could uncover just one new avenue to explore, it'd be worth the cost.

So I paid out of pocket for all that insurance wouldn't cover, adding to the list of health chores that were slowly overtaking all my time. When I got to a LabCorp for the slew of tests this functional medicine practitioner ordered, there were so many vials needed that the phlebotomist used two plastic bins to hold them all. It took nearly fifteen minutes and I felt almost as depleted as when I donated blood. For another test, I spat into tubes multiple times a day to measure my cortisol, while urine tests measured estrogen and progesterone. A new gastroenterologist also ordered a hydrogen breath test, for which I was required to drink a sugary solution and then breathe into plastic tubes every twenty minutes for three hours. This made my stomach so upset I barely ate for the rest of the day.

Even with all the many things the functional medicine practitioner tested for, the most he could say in his evaluation was that my immune system and gut were dysregulated. But if I was willing to buy several thousand dollars' worth of supplements to take on a daily basis, he was certain I'd reverse the course of my diseases, whatever they might be. I'd also need to permanently modify my diet, but plenty of his clients had done this successfully. He'd be happy to work with me.

I stared at his pleasant face on the video call, the bland minimalism of the office behind him, and briefly contemplated telling him to fuck right off. But I was so tired. I promptly ended the call, feeling as much anger and frustration with myself as with him. I had no desire to be his patient and pay thousands for supplements. I had let myself be duped by the promise of a better system, a different kind of doctor.

And as it turned out, only the test ordered by the gastroenterologist—the one that measured my breath hydrogen levels—had actionable information. I'd developed a case of small intestinal bacterial overgrowth, or SIBO (pronounced "SEE-boh," an acronym that sounded like a Star Wars character). It may have developed as a result of my food poisoning, or my Celiac disease, or some unknown cause. Gut microbes are essential for human health, as I've discussed, but they largely inhabit the colon and not the small intestine. Somehow my critters had migrated too high up the GI tract and were now causing an array of unpleasant symptoms, as well as malnutrition and malabsorption of nutrients. Both GI doctors I saw agreed that placing me on a non-absorbed antibiotic to target exclusively my small intestine would provide significant relief.

The two-week course of pills improved my symptoms to a certain extent. But there were still dozens of foods that left me nauseous, and the accumulation of extra-intestinal problems hadn't subsided either. The strange fevers, the fatigue, the swollen joints and uveitis. And because I had a wait of months before a rheumatologist could see me, I turned, again, to alternative medicine—this time in the form of a registered dietitian.

Even though we were in the same city, we initially connected over Zoom. She called from her own house, looking less like a smooth scam artist and more like someone who worked at a health food co-op. She decried the absurd prices and supplement regime recommended by the first functional medicine doctor but also insisted that her knowledge of diet and integrative nutrition had put her multiple sclerosis into permanent remission. She could absolutely help me get past my bout of SIBO, and who knew what might happen next? She believed in using Western medicine and functional medicine together; maybe I just needed the right combination of drugs and food.

I told myself, *At least she's not asking me to pay her thousands of dollars for a consultation and I am dealing with vitamin deficiencies.* I told myself, *I need the help. I don't know how to eat anymore.*

The dietitian prescribed the SIBO-specific diet, an incredibly restrictive regimen I'd need to follow for at least a month. She gave me an intimidating packet complete with a list of "legal" and "illegal" foods, organized from left to right with the allowable foods and quantities in green on the left, and the forbidden foods in red on the right. The list ran for eight pages and was highly specific. Two brussels sprouts were "legal" but six were dangerously high in FODMAPs (fermentable oligosaccharides, disaccharides, monosaccharides, and polyols). This category of carbohydrates had to be avoided to starve out any bacterial stragglers in my small intestine that survived the antibiotic treatment. I'd also need to avoid most beans and legumes, most nuts, all natural sweeteners except four types of honey, all grains, and most dairy. The dietitian gave me a website where I could look up recipes. She promised the extreme diet wouldn't last long.

I could do this. I wanted to do this, to finally feel better. Because even if I didn't quite believe her, a part of me still held out hope—maybe I'd turn food into my cure.

This is the allure complementary and alternative medicine (CAM) holds for people with chronic diseases who have few treatment options. CAM is a broad category; it includes everything from acupuncture and homeopathy to reiki and herbal supplements, all alternatives to Western medicine. And Western medicine does fail, often, because the body and the environment surrounding it are complex, nuanced systems.[8] One study of women in the UK found that they moved toward CAM practices when doctors didn't provide treatment for poorly understood ailments (like myalgic encephalomyelitis/chronic fatigue syndrome and IBS). "The more severe the imbalance or illnesses prior to CAM consumption then the more the state of vulnerability was intensely felt, which ultimately led to the most profound assimilation of the new ideology," writes researcher Kate Armstrong.[9] But those who most fully embraced CAM were also most likely to move into dangerous territory, including refusing to vaccinate their children with the safe and effective measles, mumps, and rubella vaccines.

What's more, for all their "natural" branding, herbal remedies and supplements can and do cause harmful side effects.[10] In a survey of cancer patients, more than 50 percent said they took supplements and herbal remedies, and fewer than half of these CAM users discussed that part of their treatment with the physicians providing Western treatment. Some 12 percent of the study participants were eventually warned away from their supplements by pharmacists who explained to them the risk of an adverse interaction with their treatments or harmful side effects like liver damage.[11]

Alternative medicine is a multibillion-dollar industry, and it's projected to keep growing.[12] One practitioner might be a small-business owner offering massage therapy to local clients, but then there are the multilevel marketing companies like Amway and Herbalife that bring in billions of dollars in revenue and exploit their own salespeople.[13] The problem with CAM isn't only that it is largely unregulated, but that it perpetuates a neoliberal mindset of needing to perfect one's body to continue being a productive worker ad infinitum.[14] While the goal may be to feel less sick, the route to achieving this is often through consumption of supplements or specific dietary regimes or particular exercises. The onus for improvement is on the individual, not something to be shared collectively. CAM does not solve the many problems with the US healthcare system; it perpetuates them.

Alternative therapies can be a helpful addition to Western medicine, and they can also be a scam. But we miss something fundamental if we talk about either modality outside their social and environmental contexts: namely, that the United States' legacy is one of racism, sexual violence, and genocidal displacement of Indigenous people, where access to healthcare is almost entirely dependent on having health insurance. More than forty-seven million Americans were affected by food insecurity in 2023.[15] In Chicago, where I live, the situation is even more dire. Latinx communities in Chicago have rates of food insecurity around 29 percent, and for Black communities it's even higher, at 37 percent; for point of comparison, 13.5 percent of households nationally are food insecure.[16] When someone with diabetes doesn't have access to food, they can easily end up in the hospital with abnormal blood glucose levels. And the communities most likely to struggle with access to food are those who have a higher incidence of chronic illnesses.[17]

Health equity researcher Arline Geronimus coined the term "weathering" for the interconnected social, political, and economic circumstances that lead to worse health within marginalized communities. "I call this process 'weathering' because that word is a contronym: Weathering can be a sign of deterioration and erosion as in 'the rock was weathering'; and weathering can also be the opposite: A sign of strength and endurance as in 'the family is weathering the recession.' For health and aging, it can be both," Geronimus writes, adding, "To be weathered is to be subjected to the structural challenges and existential insults that our society creates for those we marginalize. Weathering also refers to how oppressed groups strategically organize family economic and caretaking systems to resist such erosion, and how these tenacious and hopeful efforts, while, on balance, protective, can also exact health costs."[18]

At my lowest, I felt on the verge of disintegration from my illnesses. But I still had the money to buy and physical ability to cook fresh vegetables and unprocessed meat in order to follow the SIBO-specific diet. On days when I was too tired to cook, my partner would do it instead. Eating nutritious food that fits whatever dietary restrictions one must follow should not be contingent on money, access, time, or cooking ability. But it is now—and it may get worse in the near future.

There are some things we can predict.

Climate change and our agricultural system are interacting systems; higher temperatures, rising CO_2 levels, changing patterns of rainfall and aridity will all affect the amount of food produced as well as its nutrients levels. But the global food system as it exists today contributes about a third of greenhouse gas emissions to the atmosphere.[19] This includes everything from food production to distribution and waste, so perhaps it's not a surprise. And maybe it's not a surprise that livestock and the food they eat account for more than 50 percent of those emissions. That's partly thanks to the microbes living within cows' rumens, the portion of their digestive tract that breaks down plant fibers into digestible fatty acids. Those microbes also happen to produce methane in the fermentation process, which is then burped out by the cows. (Okay, yes, a small amount is expelled as farts.)

About 40 percent of the planet's surface is currently being used for agriculture, but if we don't increase global food production by 70 percent from their 2005 levels by 2050, there will not be enough to feed the projected 9.1 billion people inhabiting the planet.[20] But we'll be fighting the effects of climate change while we go about this process; in the worst-case scenario where emissions increase and we go whole-hog on fossil fuels, some one-third of currently arable farmland will no longer be viable for agriculture.[21] Already, between 720 and 811 million people are undernourished, and 2.3 billion people are affected by malnutrition. Tens of millions more people could be facing food insecurity by 2050 if we fail to curb carbon emissions. Everything from fortifying supply chains to creating farms with more crop diversity and better soil health will play a role.

There are other things currently beyond the scope of our science.

On February 7, 2021, 26.9 million cubic meters of rock and ice broke away from the north face of Ronti Peak, a 6,603-meter cliff within the Indian Himalayas.[22] Though 80 percent of the mass was made of rock, the 20 percent of ice quickly liquefied from the heat generated from friction and sent the whole mass of material careening down valleys and straight into two hydropower projects, the Rishiganga and the Tapovan Vishnugad. Of the 204 people killed or missing in the landslide, 190 were workers at the two sites. Known as the Chamoli disaster, the amount of sediment and debris was so staggering that it affected water quality in the Ganges River some nine hundred kilometers away, leading to a level of suspended sediment over eighty times higher than the permissible limit in Delhi. While scientists know that climate change is causing rapid glacial melting and mountain slope failures, it's still unclear what triggered the Chamoli landslide and how much of a role climate change played.[23] Why did it happen that day, and not another?

Why did the Oso landslide happen on a Saturday instead of a Monday, when more people would've been out of their houses? Smaller landslides had happened on the slope in decades past, most recently in 2006, shortly after which the state of Washington built a crib wall to retain sediment.[24] Figuring out why any landslide happens when it does is a challenging question to answer because, as geologist Dan Shugar explains, mountains are always moving. They grow and they shrink from erosion, and much of the movement happens on a slow, slow timescale—but sometimes it happens quickly. Catastrophically. Sometimes scientists can identify the underlying cause (warming causes ice to melt and rocks to become unstable) but not the trigger that was the final factor in a collapse. "Mountainsides tend to be pretty opaque things and we can't really easily see inside them," Shugar says.[25]

Scientists are attempting to develop satellite technology that might help predict catastrophic landslides in advance, to give people a chance to flee their path.[26] We have satellites that can take color photos of the same location over time and show boulders moving, LIDAR that builds three-dimensional models of topography. Even satellites that can accurately measure down to the millimeter from space, Shugar says, though they come with the disadvantage of the image getting obliterated if the movement is too large. "We've got wonderful tools to monitor slopes from the air or from space," says Shugar. "The challenge is we have even more slopes to look at, and at our current technological state, we don't have the tools to do this in an automated way for vast swaths of mountains."

For now, survivors of landslides and the family members of those killed have to learn how to live with their losses. In a March 2024 documentary on the ten-year anniversary of the Oso slide, multiple community members described the horrors of recovering bodies and learning of the deaths of loved ones.[27]

"I remember every single detail of it like it was yesterday. I heard a rumble and so then I looked out our front door and our neighbor's chimney was coming straight at us and all I could do was turn," said Amanda Suddarth, who was with her newborn son, Duke, at the time of the slide. They both lived through the experience, though Duke, who was around ten when the documentary was filmed, has daily seizures from the traumatic brain injury he experienced.

"You couldn't let yourself get tired because you have to keep going, because we were gonna find everybody," said Jan McClelland, a volunteer firefighter from Darrington who was on the scene for months, excavating remains.

"I couldn't speak her name, I couldn't talk about her for five years, cause it hurt so bad," said Seth Jefferds, who lost his wife and granddaughter in the disaster. "Ten years later it still hurts."

"You would think that time and distance would help, and the thing is, it's never gonna help, because I've seen the big and the bad and the ugly in my career; I've seen everything, everything you can imagine. And I had not seen that," said John Pennington, former director of emergency management in Snohomish County, Washington.

"It's like there's a scar there, it's always gonna be a visual for us," said McClelland. "And honestly, I always think of the people every time we drive through there."

Between 2016 and 2018, multiple families and survivors brought lawsuits against the state of Washington and the Grandy Lake Forest Associates timber company, alleging that the crib wall made the leading edge of the landslide deadlier than it might otherwise have been. Ultimately, the state and the timber company settled with the plaintiffs, paying around $71.5 million to the victims' families and survivors.[28] In January 2021, the National Landslide Preparedness Act was signed into law. It aimed to improve our national understanding of landslide risk and develop mitigation measures. It's a move in the right direction, and it doesn't bring back anyone who has died in landslides over the past decade.

Some foods will become harder to grow; landslides will happen more frequently. It is not hyperbole to say that the severity of repercussions depends on how quickly we move our economies off fossil fuels. To prevent the very worst outcomes, change must happen as soon as possible. If we drag our feet, if fossil fuel corporations put all their energy into greenwashing while still racking up permits for new drilling leases, millions more people will die.[29] Even if we act with all haste, the planet has already been transformed. There is no going back.

Sometimes the grief of this knowledge ripples out from me. Or maybe it's rippling out everywhere, like the seismic waves produced by undersea landslides, and my body rings with it like a hollow bell. I feel immobilized by those vibrations, hearing nothing but the words "it's too late," over and over.

And sometimes, grief is itself a landslide that obliterates a known landscape, forcing its topography to become something else entirely. When a high school friend suddenly died while I was in college, I felt as if my inner compass were spinning endless circles, not pointing anywhere. We hadn't been close friends, but everyone knows everyone when your graduating class is just over 150 people. I knew Jessie's family. We'd been on the same swim team. I just couldn't fathom it. (The same year as Jessie's death, researchers spotted the last known member of the Christmas Island pipistrelle bat; the species is now considered extinct.)[30]

One of my closer friends died just as suddenly when I was in grad school. I'd been to so many parties at Angie's godfather's house, spent the night there, visited her at her job scooping ice cream. We always paired together in gym class because neither of us liked running. We talked shit about our teachers. We were both in the school plays. We were weird and goofy and disagreed on things sometimes, but I thought she'd always be there, someone to meet up with when we were both back home on school breaks. Then one day she got sick with strep throat, the infection worsened, and she died. (The same year as Angie's death, a tiny catfish found only in Ohio called the Scioto madtom was declared extinct.)[31]

In the long winter months at the end of 2019 and the early part of 2020, when I was in the midst of my unnamed illnesses, I wondered if I might die the way my friends had: suddenly, too young. This didn't stir up the same level of fear and anxiety as my heart palpitations did, but instead laid me low with a heavy, unending grief. I just wanted to go back to who I was, to my old body. I wanted to stop being in pain. I wanted to be able to eat food again, and not by following such an extreme diet that my therapist began worrying I'd develop the eating disorder orthorexia.

It was on one of those endless nights of exhaustion, depression, anger, and confusion that I found Eli Clare's book, *Brilliant Imperfection: Grappling with Cure*. Clare describes not only his experience with cerebral palsy but also with gender transition, with sexual assault, with the history of eugenics in the United States. The book by turns made me furious and hopeful. Early on, he describes strangers wanting to pray over him or offer crystals, all with the goal of curing him. "I have no idea who I'd be without my tremoring and tense muscles, slurring tongue," Clare writes. "They assume me unnatural, want to make me normal, take for granted the need and desire for cure."[32]

And still I didn't understand what he meant when he wrote that cure justifies violence and oppression.[33] How could he possibly argue that quadriplegic Christopher Reeve searching for a cure to his quadriplegia was spreading propaganda through a documentary called *Christopher Reeve: Hope in Motion*? But the deeper I got into the book, the more my heart cracked open with a new kind of rage and grief. Because attempted cures *are* sometimes violent (as when autistic people have been sprayed with ammonia or restrained in chairs).[34] Cure is not as straightforward as the many,

many research foundations make it sound when they're advocating a cure for cancer, or Alzheimer's, or any other disease—even when those diseases are terrible, and treatment is lifesaving.

I didn't know yet if I could call myself disabled. I'd been sick for most of my adult life, and it was only getting worse. But most days I could walk, and most days I could shower, and most days I felt half-alive. What did that make me? Even if I didn't yet have the confidence to don this identity, after reading Clare's book I knew I had to let myself breathe into the possibility of not finding cure. I needed to stop chasing impossible diets and supplements. I needed to stop trying to return to a bodymind that no longer existed.

When I made the choice to walk away from the abled ideal of "cure," I understood I was also opening myself up to more profound and regular grief. If I believed in the work of supporting chronically ill and disabled people wherever they were at—of counting myself among them—it required being clear-eyed about who might die. A disability activist I followed on Twitter posted about getting ready for a surgery and being scared but wanting good books to read for recovery. She asked if people would send them to her. I picked one from her list that I'd also read and loved and mailed it over, wishing her the best for her procedure. When I checked in again, I learned she'd died. And I didn't really know her at all, but I grieved, because disabled and chronically ill people—especially those marginalized in other ways—die too young. It made me think of the poem "Kindness" by Naomi Shihab Nye. "Before you know kindness as the deepest thing inside, / you must know sorrow as the other deepest thing. / You must wake up with sorrow."[35]

My whiteness, my affluence, even my health insurance had insulated me from this firsthand knowledge for years. I was now so sick with so many conditions that my privileges could no longer protect me. But that terrible realization was, in a strange way, emancipatory. Like seeing the code that powers the Matrix instead of the illusion of reality, I was beginning to understand that we live under an economic and political regime where the disablement of humans and the environment is built into the system.[36]

The lessons came at a critical moment: Earth's temperature was rising fast, a virus called SARS-CoV-2 started circulating around the world, and I was still the sickest I'd ever been. A lot more people were about to get sick and lose their homes, and too many were going to die.

PART FIVE

FIRE—JOINTS—RADICAL LOVE

BENT BODIES AND HIDDEN FIRE

In my childhood, fire had two faces.

The first: friendly, vivacious, exuberant. Like a puppy with sharp teeth and too much energy, capable of doing damage but not out of malicious intent. Fire as community, as celebration, as connection. Bonfires on the beach made of sand-crusted driftwood, pulsing an unpredictable pattern of light and heat into the starry night, sending glimmers off the black lake. Pine-scented fires made from dried-out Christmas trees, around which my cousins and aunts and uncles and friends huddled, the flames sometimes erupting blue as a forgotten strand of tinsel burned up, nearby ice hissing as it melted. Campfires to cook hot dogs or roast vegetables wrapped in foil or toast marshmallows to the perfect shade of golden brown. Simmering prairie burns that leave the earth blackened, amplifying the sun-fed warmth to spring plants, smelling like sweet charcoal. I loved all the fires, always volunteered to get them started, wanted nothing more than to sit and watch the ripple and dance of flame till there was nothing but snapping embers.

The second: deadly, unpredictable, bad to know. House fires that killed sleeping children or leveled historic cities. Grease fires in the kitchen, subverting the normal knowledge of adding water. Frayed wires allowing electrons to escape their confines and ignite furniture. Stop drop and roll, in emergencies break glass, test the smoke detector on a regular basis. In case of such an alarm going off, feel the doorknob before opening your bedroom door. Block smoke with a towel. Open the window, hang from the ledge, and drop if you have no other option. Better to break an ankle than burn to death or die of smoke inhalation. Fire has no place indoors, except on carefully monitored candles, and the friendliness of a bonfire should not be mistaken for house-training.

Wildfires belong to this second category as well, but on a far more massive scale. They were also, notably, not a threat to my particular corner of North America; the Chicago fire of 1871 was old history by the time I moved there, even if it is emblazoned on the city's flag in the form of a star. We had heat waves and polar vortices, blizzards and tornadoes, floods and droughts. I took comfort in the knowledge that fire happened in different ecological regimes.

This was, of course, some mixture of ignorance and hubris. Not only does the Chicago area have its own fire weather, but the planet shares one atmosphere, and all that travels by air makes its way across the globe in varying concentrations. The older I got, the more I understood I was not immune to wildfires. I worried for West Coast friends as fire seasons grew longer and more explosive.[1] I spoke with forestry experts who monitored the fires on palm oil plantations in Indonesia, read about flames threatening endangered orangutans.[2] Wildfires in the boreal forests of Eurasia and North America have tripled since 2001, leading to a 60 percent increase in forest fire carbon emissions.[3] Even pop culture reflected the world's out-of-control fire landscape: a cartoon of a dog in a miniature bowler hat, sitting by his mug of coffee in a room aflame, asserting, "This is fine," went viral.[4]

Starting in mid-April 2023 and continuing through October, Canada experienced its worst wildfire season on record, with more than 18.4 million hectares burned (an area slightly larger than the size of North Dakota).[5] The forest fires were more than seven times larger than the average burned area recorded over the forty years preceding them, which was attributed to the fact that 2023 was also the warmest and driest year in that forty-year period.[6] More than six thousand fires burned across Canada, many triggered by lightning storms, smoldering away in remote locations across British Columbia and Quebec and the Yukon.[7]

As conflagrations erupted, smoke and particulate matter flowed down over vast swaths of the United States.[8] In New York, the sky turned an apocalyptic shade of orange. In the Great Lakes, satellite imagery showed a gray haze so thick it obscured most of Lake Michigan and the state of Michigan.[9] The spike in air pollution was so extreme and anomalous that one day in June, media outlets rushed to proclaim Chicago had the "worst air quality" of any major city.[10] People were warned to stay indoors, or wear masks outdoors to protect their lungs against the dangerous particulate matter, due to the heightened risk of asthma attacks and lung damage. On my own brief foray outside the house, the air stung my eyes and seemed to clog my throat even with protection from an N95 respirator. An acrid smell hung on the weave of my clothing when I returned indoors.

It was the fourth year of a deadly pandemic, and nearly 7 million people had died globally of COVID.[11] Just that May, the Biden administration ended the federal Public Health Emergency, terminating, among other things, authorizations by the Centers for Disease Control and Prevention (CDC) to collect public health data about cases of the virus, the free availability of at-home COVID tests, and free access to treatments like the antiviral Paxlovid.[12]

I wondered, breathing the noxious smoky air through my mask, if any part of my difficulty breathing came from my first bout of COVID several months earlier. Without an at-home test, I wouldn't have known so early that I was dealing with the virus. Without access to the antiviral Paxlovid, I likely would've ended up in the hospital. Even with the medicine, it was a close thing. A ragingly high fever combined with extreme body aches, incessant vomiting, a terrible cough, and no appetite. I was barely consuming any food—and, at times, it seemed, barely any oxygen. The worst passed in a couple weeks, but the cough lingered, turning into bronchitis. Fearing I'd develop pneumonia, my doctors placed me first on antibiotics, then an inhaler, then oral steroids.

I'd received every updated COVID vaccine, often at double the "normal" dose due to my status as an immunocompromised person, and this was still how COVID presented itself in my body. The virus was another fire on top of my body's own abnormal inflammatory response, the autoimmunity that had settled into my joints and triggered itchy, bleeding rashes on my scalp. The official diagnosis of psoriatic arthritis came only after my bout with COVID, possibly because my body went for months without any immunosuppressant medication to dampen the effects of my runaway immune system. By the summer of 2023, when the smoke of Canadian fires made its way down to us, I was tired and achy and itchy and scared.

But the drama of wildfire smoke overshadows a more insidious reality: the air quality in Chicago has long been detrimental to the health of its residents. In 2020, researchers estimated that air pollution from vehicle traffic and industrial activity contributed to 5 percent of premature deaths in the city.[13] As of 2024, Cook County (which includes Chicago and multiple suburban cities) was in violation of EPA air standards, with especially bad ozone and particle pollution. And it's not the only place where air quality is hazardous or unhealthy: the American Lung Association found that more than 131 million people live in places that have received a failing grade for at least one measure of air pollution.[14] When we think of "fire," most people tend to forget that internal combustion engines are also burning, that coal plants are burning, that natural gas is burning.

Not Stephen J. Pyne. The fire researcher coined the term "Pyrocene" to describe our current era of mega-wildfires and fossil fuel–driven climate change, a fire age instead of the ice ages that dominated the earlier history of *Homo sapiens*. He envisions a trio of fires on the planet. "First-fire" is nature's fire, triggered by lightning strikes and volcanoes and even rockslides generating sparks. "Second-fire" is humankind's initial relationship with fire, for food and warmth and landscape transformation. "Third-fire" comes from lithic landscapes, the coal and oil and gas that were buried for eons until humans learned to harness their raw power—without necessarily understanding the ways this would undo the world.[15]

"This was geology's version of *Jurassic Park*," Pyne says of fossil fuels, an analogy that came to him after watching one of the most recent films in the franchise and seeing dinosaurs surrounded by fire. "You've taken stuff out of deep time, tens of millions of years ago, and you've brought it to the present and released it. And it has no place to go, it doesn't belong here. It's not just energy from burning the fossil fuels, but all the other biomass, the petrochemicals, the plastics, the tar, the asphalt."[16]

Pyne, a self-described "pyromantic," has worked with fire for the entirety of his career, beginning at age eighteen on a fire crew at Grand Canyon National Park. He's studied the fire regimes on every continent, learning not only from the fossil record and decades of policy on fire suppression but also Indigenous and Aboriginal fire practices. "European elites treated European peasants with the same disdain with regard to fire [as they did with Indigenous people]," Pyne says. "Most European agriculture relied on fire in some form. But it was just dismissed as primitive and superstitious. You've got people in the late nineteenth century saying that if you used fire, you are backward, and if you find an alternative to burning, you are Rational with a capital R."

More than mere prejudices, these were the attitudes on which colonial empires were founded. Alongside the exploitation of natural resources and humans of all kinds was the need for control in its many forms.[17] Control of nature, control of bodies. As environmental historian Jason W. Moore writes, "Capitalism was built on excluding most *humans* from Humanity—Indigenous peoples, enslaved Africans, nearly all women, and even many white-skinned men (Slavs, Jews, the Irish). From the perspective of imperial administrators, merchants, planters, and *conquistadores*, these humans were not Human at all. They were regarded as part of Nature, along with trees and soils and rivers—and treated accordingly."[18]

We know of no other planet in the universe that contains all the elements for fire. Until we find life on other planets, we're unlikely to find fire. And that's because it isn't enough to have lightning; fire requires fuel, and that fuel comes from life. On Earth, the oldest fossil charcoal dates to around 420 million years ago, relatively shortly (in geologic time) after the appearance of the first land plants.[19] This interplay has continued ever since, with some ecosystems adapting specifically around fire in the same way others evolve around an abundance of rain. Pyrophytic plants like lodgepole pine and Eucalyptus trees have resinous cones that require fire to unseal their seeds. In fire-prone parts of South Africa, smoke can trigger germination in some plants. Multiple species of larch tree in the boreal forests of Siberia rely on fire to open the terrain for seedlings and limit the growth of evergreen conifers. Fire is destructive, but it is also generative; flame removes pests, burning resets the board for new life to take root, ash fertilizes the soil.[20]

Surely ancient hominins saw the way fire shaped the growth of new landscapes, what happened when plants and animals burned. At some point they made the connection between the heat and light of fire and its potential as a tool. Archaeologists hypothesize that by 1.7 million years ago, when *Homo erectus* was expanding beyond Africa, these primordial cousins of ours may already have had some knowledge of fire.[21] Their smaller teeth suggest access to cooked food; when meat and tubers are cooked, they're easier to digest and more calorically dense.[22] By 1 million years ago, hominins were using fire within caves, indicating they didn't simply wait for lightning strikes to ignite the landscape—they'd learned how to transport fire.[23] Starting around 420,000 years ago, Qesem Cave in modern-day Israel shows evidence of habitation that lasted more than 200,000 years, and fire was present for that entire duration.[24] Charred bones, butchering tools, and even microscopic remnants of charcoal on fossilized teeth all indicate the extent to which fire was being used for cooking meat. Fire shaped the lives of our hominin ancestors. As Pyne says, it could be considered humanity's first act of domestication.

But mastery of fire is only one component of ancient populations that modern humans likely recognize and relate to; another is the presence of care in community interactions. Contrary to the image of ancient life as a violent "survival of the fittest," there is evidence for community members supporting the disabled long after these individuals offered any "utility" to the group in terms of hunting and gathering. Fossilized remains of a Neanderthal child indicate she had severe inner ear deformity, likely as a result of Down syndrome, which would've caused extensive hearing loss and balance issues.[25] But the child lived past the age of six, a feat that likely would not have been possible without support from the mother and other community members. Another individual, this one from around eleven thousand years ago, was discovered to have acromesomelic dysplasia, a form of dwarfism that resulted in growth deficiency and restricted mobility at the elbows.[26] The condition would've made it much harder to hunt and move over the landscape. "Yet despite these handicaps," the researchers write, "the [individual] survived to about 17 years of age. Moreover, once he died, he was accorded special funereal treatment as evidenced by his burial in an important cave."[27]

And around four thousand years ago in northern Vietnam, a young man was paralyzed from the waist down with very limited upper body mobility, yet he survived for a decade thanks to intensive community care.[28] He likely required assistance eating and drinking, as well as for personal hygiene and transportation. And he *received* this care, during an era that we associate with minimal technology and brutal calculations about life and death.

Archaeologist Lorna Tilley has coined the term "bioarchaeology of care" and created a framework to help other researchers analyze the remains of sick and disabled individuals for the level of human intervention that would've been required to support them.[29] Her background in healthcare before transitioning to archaeology was the impetus for developing this framework.[30] As she writes, "Research into disability seeks access to experiences and behaviors which, in a living individual, are always personal, often powerful, and sometimes deeply private. While the human remains at the center of study no longer retain sentient existence, the research process itself endows them with *social* existence."[31] Tilley notes that bringing such an approach to archaeology can help combat the undervaluing of caregivers, help us recontextualize the past, and emphasize the existence of sick and disabled people throughout history.

Tilley's work reminds me of the blog by disability justice organizer Mia Mingus. "We must leave evidence," Mingus writes. "Evidence that we were here, that we existed, that we survived and loved and ached. . . . Evidence for each other that there are other ways to live—past survival; past isolation."[32] The blog itself is titled "Leaving Evidence."

I want to believe in the infinite capacity of humans to treat the disabled with acceptance and respect. Not as disposable, not as burdens, but as complex human beings who have every right to live satisfying lives. I want an end to the cultural narrative that says it's better to be dead than disabled.[33] I want to look into the deep past as a source of inspiration for how we might navigate the uncertain future, because in that past I see something like love. If hominids living on an Ice Age landscape filled with diseases and food scarcity and threats from saber-toothed cats and other megafauna were able to care for their disabled kin, why shouldn't we, in our age of fire, with our immense advances in technology, be able to do the same? There will always be threats of one sort or another, and chores and disagreements and little hurts. We can't know for certain that love was the motivation for those ancient people who helped their kin survive disablement. But it's what I choose to believe. It's how I want to frame my own relationships: that we all take risks for the betterment of each other, because we create a better world when all of us are thriving.

I want to believe we can make this choice, because so often in the more recent past, we have moved toward its opposite. We have done everything in our power to hide, sterilize, or kill anyone whose bodymind was determined deviant.

Revolution seemed everywhere in the eighteenth century, from the halls of power to all corners of natural philosophy—the soon-to-be-modern sciences. British ironmonger Thomas Newcomen learned of the challenges in pumping water out of tin mines and turned his mind toward solutions. He invented the Newcomen engine, sometimes called a fire engine, to pump water from both tin and coal mines.[34] Before the end of the century, James Watt patented the steam engine, which improved on the up-down movement of Newcomen's piston-driven engine and instead produced continuous circular motion. This meant the machine had applications far beyond pumping: it could power distilleries and replace water mills in the cotton industry. It moved centers for manufacturing away from waterways and into densely populated urban areas. The only constraint on these new machines was the need for laborers and coal. As Watt's business partner Matthew Boulton would say of the steam engine, "I sell here . . . what all the world desires to have—POWER."[35]

Fire wasn't just being forced into machines; it was also being pulled apart in the chemist's laboratory. In 1778, Antoine Lavoisier proposed the role of oxygen in greatly varied processes, such as rusting and burning. Fire was not an elemental force but a calculable process of combustion, something to be measured and manipulated, studied and contained. With the newly invented coal-powered "fire machines" pumping water out of mines, there was an ever-greater source of lithic fuel available to power the Industrial Revolution. By 1850, Britain alone was responsible for more than 60 percent of global CO_2 emissions and excavated three and a half times as much coal as France, Germany, Belgium, Austro-Hungary, and the United States combined.[36]

As went mechanical engineering, chemistry, and mining, so went medicine. Physicians began moving away from the four humors that had reigned for millennia. No more would blood, black bile, yellow bile, and phlegm dictate the health and personality of individuals (and how their bodies responded to diet and weather). Medical science divided itself again and again into specialties, penetrating the body ever further to understand how disease wreaked havoc within tissues and organs.

In 1741, a French physician invented a term that would from then onward steer a developing practice. In a work titled *Orthopedia: Or, the Art of Correcting and Preventing Deformities in Children,* Nicolas Andry instructed parents and caregivers on how to identify early signs of crippling and forestall them.[37] The term came from Greek: *orthos* for "straight and free from deformity" and *paidios* for "child." As if to emphasize his point, Andry included an illustration of a crooked tree tied to a straight post. He also provided instructions on what to do if a child was developing bowlegs as they began to walk: either prevent them from walking until their muscles were stronger, or, if it was already too late, apply "a small plate of iron upon the hollow side of the leg" and fasten it there, tightening it every day until the leg is sufficiently straight, writes surgeon and medical historian Leonard Peltier.[38] Apparently the only thing needed to prevent pain from this treatment was some padding between the plate and the child's leg.

As the name suggests, the focus in the early years of this field was treating children, either by surgical means or with restraints (a brace for treating clubfoot, and for scoliosis a large contraption to sleep in, both invented by Swiss orthopedist Jean-Andrè Venel).[39] Aside from congenital disabilities, there were plenty of diseases that could inflict musculoskeletal damage in young children. Polio could cause paralysis; skeletal tuberculosis could trigger spinal deformities and abscesses as well as muscle weakness and joint inflammation. To repair the damage done by these diseases, orthopedists relied on braces, massage, and ever-changing surgical techniques. In the United States, an orthopedic surgeon who opened a practice for "crippled children" later applied his skills to soldiers injured in the Civil War. Elsewhere, such as in the UK, the trajectory was similar. As medical historian Anne Borsay writes of the post–World War I era, "On return to civilian life, orthopedic surgeons advocated 'an ambitious scheme to recreate . . . the glory of their military empire' by organizing a national program for the 'crippled' children with whom they had started to work before the war."[40]

These weren't the only ways in which the bodies of children were co-opted for larger agendas. American polio survivor Blanche Van Leuven Browne spent much of her childhood in a hospital following her paralysis by polio, and her experiences led her to create her own institution for "crippled" children that emphasized their education and centered the children as knowledge holders about their bodies. (It should be noted that this hospital and others affiliated with the "crippled child movement" of early twentieth-century disability activists were overwhelmingly white and segregated, even in northern states.)[41] Although Browne had collegial relationships with doctors, she was eventually pushed out of the institution by funders who wanted a more medicalized approach to treatment and cure.[42]

In response, Browne penned a pamphlet titled "On the Vivi-Section of Crippled Children" that emphasized the fact that many orthopedic surgeries were not only unnecessary but also experimental. Historian Leanna Darlene Duncan writes, "It was not uncommon for parents to give up on a child with a disability, convinced that they would never be cured enough to live a useful life or perhaps that they would die within a short time. Hence, it was easy for these doctors to gain the consent of desperate parents or distant state authorities to try something that promised a chance at recovery, even if risky."[43] Finding patients on whom to attempt new orthopedic surgeries was even easier when doctors went to orphanages, where few people would speak on behalf of the child.

Power concentrated at the top of the medical profession, despite patients attempting to push back. It's a pattern that continued throughout the twentieth century, even as the disability rights movement grew. Researchers Jeanne Hayes and Elizabeth "Lisa" Hannold point to the work of French philosopher Michel Foucault on the origin of modern prisons to shed light on the ways that surveillance and control extended beyond the prison system and into all fields, including medicine. "These new forms of disciplinary power," they write, "recast the body as an object that could be controlled, improved, and transformed."[44] Foucault even repurposed the famous bent tree image associated with orthopedics as a visual representation of how physical, social, and emotional deformity were thought to require policing and correction.

One example of this ethos is Diane Fields DeVries, born in 1950 with no legs below the hip and arms that began above the upper elbow. From early childhood she was a patient at UCLA's Child Amputee Prosthetics Project, funded by federal grants in a period "characterized by an ideology of rehabilitation and faith in technology."[45] Though she learned to wear prostheses on her arms and legs from age four to eighteen, DeVries ultimately rejected all of them; it was much easier and more natural for her to use her stumps for tasks like reading, writing, and holding a cup. When she began stating her preference for dresses with narrow straps that allowed her the most freedom of movement, hospital staff expressed extreme discomfort. In one note, they wrote, "Those present felt that this is somewhat unattractive and possibly disturbing," writes researcher Gelya Frank.[46] The doctors framed prostheses as a matter of dependence versus independence to DeVries's parents, stating she would likely be institutionalized for the purposes of daily care if she refused to make use of prostheses. Their dire warnings did not, in fact, come true.

Researcher Lisa Hannold described a similar experience with orthopedic doctors, who wanted her parents to consent to a surgery for "drop foot" caused by spinal muscular atrophy. The surgery, in combination with bilateral leg braces and orthopedic shoes, wouldn't give her the ability to walk, but she would have "a more normal looking foot and the ability to wear 'pretty shoes' as she grew older."[47] Hannold was relieved when her parents refused.

I'm reminded, too, of Audre Lorde's experience with a mastectomy to treat her breast cancer, and her refusal to wear a prosthetic or undergo reconstructive surgery to appear more "normal," which she describes in *The Cancer Journals*. In another essay in the book, called "The Transformation of Silence into Language and Action," Lorde asks, "What are the tyrannies you swallow day by day and attempt to make your own, until you will sicken and die of them, still in silence?"[48]

There's a moment before surgery begins that you may sometimes be dimly aware of: the crossing of a liminal space that transforms you from *person* into *body*. It's not when the surgeon visits your room to initial the appendage slated for the knife; he wants your initials, too, letting you mark yourself to confirm you understand what is about to be cut open and stitched shut. In this moment you're a team, patient and doctor. It's not when the anesthesiologist visits, either, though they may be brusque as they note the shape of your mouth and teeth, jot down any reactions you've had to anesthesia in the past. (We're still learning the exact mechanisms for how general anesthesia works on the brain. It's always seemed like some version of the underworld to me. I tend to wake on the border of dreams, dry-mouthed and shaking, as if my body holds onto the memory of what happened in the surgery even though my pain receptors were dulled.)

The transition isn't even when the nurse wheels you to the OR and pumps an anxiolytic into your bloodstream, the Versed or Halcion or Restoril that makes the room spin. It's right after that: watch. Groggy, half-awake, you move yourself from the padded hospital bed with its blankets and sheets onto a much harder slab, where your limbs and torso and head will be manipulated into position by others. That scene, glimpsed briefly, is when you become *body*.

I had a nightmare once, around the time of my hip surgery (the third of my joint surgeries so far), of seeing myself from the outside as the nurses lifted me off the operating table while I was still unconscious. In the dream, my legs were splayed and bloody, my head lolling against a stranger's chest. I awoke to something between a scream and a sob.

Arthroscopic surgery is a medical miracle, a way of penetrating the body's barriers with the smallest of incisions, narrow tools with cameras and lights and razors and needles allowing a surgeon to see within from outside, without splitting the body wide apart. Recovery from arthroscopy tends to be faster than open surgery, the risk of infection lower. But those little slits of sundered skin and muscle bely the trauma that has still happened. For my knee surgery, filling the joint with water so that the skin swells, melon-like, leaving room to maneuver in the narrow space between patella, femur, and tibia. For my wrist surgery, the bone cutting that was slated to happen but did not: slivers of ulna—a living organ—hewn away. For my hip surgery, the joint was dislocated, leg hung at an uncanny angle so the scopes could probe the concave socket and trim frayed tissue away, sew together torn cartilage and anchor it with five nails into the bone. For all of them, a tube down the throat to perform the essential function of breathing. Blood washed away from open wounds. The nakedness of nothing under the hospital gown, but also the nakedness of being unaware for all of it, as a suite of strangers dissect your body like it's any other Wednesday, because for them it is; for them you are one surgery of many, one body among hundreds, flesh to be opened and repaired and closed shut again, like a car at the mechanic's.

Here are the contradictions of care in orthopedics. The surgeons are present for a fraction of the process, between the operation itself and perfunctory follow-up visits that sometimes require sitting in the waiting room for two hours. Yet these doctors are paid a median salary of $724,000 per year in Chicago, six times higher than the average physical or occupational therapist with whom a patient will spend months in rehabilitation. There are the nurses, sometimes so full of kindness it causes tears: locating gluten-free snacks, holding a cup of ice cubes for you to chew as you battle nausea, pushing your wheelchair all the way to the parking garage. There are other nurses who might be cruel, who refuse a request to wear a mask for an immunocompromised patient, who don't apologize for blowing out a vein and allowing a golf ball–sized globule of stinging fluid to collect under the skin. There are the scheduling staff who orchestrate this great flow of humans between office and surgical suites, the billing department that may do battle with insurance agencies or turn their pillaging toward the patient.

And there are the spouses and friends and family who cook and clean and send flowers and cards and cookies, who carry glasses of water from one room to another, who help with showers and getting dressed and using the toilet. (For weeks after my hip surgery, my partner wrangled a sock onto my right foot because I couldn't bend to do it myself.) This is the unpaid, unseen, uncelebrated care work that I imagine from the archaeological record. This is what I picture instead of a bent tree being forcibly straightened: a forest of trees burnt by wildfire, still alive below ground, collecting nutrients with the help of their mycorrhizal network.[49] I think of Blanche Browne and her pamphlet on the vivisection of children, how so much of medical history preserves the names of surgeons who pioneered new techniques but not those of the children pulled from orphanages to be operated on because their bodies did not conform.

And maybe some of those children found relief from pain in surgery, and were grateful, as I am. But who cared for them during the long recovery? Where did they turn for comfort as they reacquainted themselves with their bodies? Could they seek restitution if things went wrong—if they wanted to return to what had been? How many of them died? Were they happier with the shape of their body before it was cut into and reshaped? I want all those marginalized by the shape of their bodies at birth, or their abilities, or disease, to teach me how to fight for change. Because, as Brazilian philosopher Paulo Freire wrote in his seminal text *Pedagogy of the Oppressed,* the oppressed understand the need for liberation better than anyone. "And this fight, because of the purpose given it by the oppressed, will actually constitute an act of love opposing the lovelessness which lies at the heart of the oppressors' violence, lovelessness even when clothed in false generosity."[50]

I've seen pity and false generosity in the faces of doctors, in narratives about trying to cure or improve someone with disabilities. In trying to understand the illnesses in my body, I've looked to history for knowledge and solace. What I've found are horrible, hopeful, complicated stories. The line between help and harm can be so thin it turns transparent.

I've learned, too, from my own life. I have been to the underworld seven times now, seen many ways that guides offer assistance or present obstacles, and I can say: our culture is not one that values care. Our society values capital, and humans as capital, and fixing what is broken so it can produce more capital. And maybe this system is part of what has led to the invention of miraculous tools and resuscitations from death, but it has also made us forget how to love the animal bodymind, the soft creature that craves to give and receive care for no reward but the act itself.

COLONIAL CONTROL

For a time, humankind and fire were inseparable. Where the one went, the other followed. Imagine how it must have looked far above the planet: where today our astronauts and satellites look down on a web of electric lights spattered across the darkened landscape of the planet in its turn away from the sun, several thousand years ago it was a network of small fires, burning in every inhabited region. When Europeans began colonizing Australia, explorers who moved from the southern coastal towns toward the Northern Territory noted how frequently the Aboriginal groups set fires. Phillip Parker King, a navigator who surveyed Australia's coastline, described the scene in 1827: "The fires which had been lighted in the course of the day by the natives, had rapidly spread over the summit of the hills, and at night, the whole island was illuminated, and presented a most grand and imposing appearance."[1]

Several decades later, John McKinlay was tasked with finding a settlement site in the north and recorded his own fire observations in May 1866. "In every direction in the distance from westward to the north we see daily innumerable bush-fires, showing the whereabouts of the natives, who must be numerous. We occasionally see recent traces of them even on the tops of these rugged walls, where they have been firing for the purposes of getting wallaby," he wrote.[2]

This type of burning happened around the world, from sub-Saharan Africa to South America. Savannas, the most fire-prone environments in the world, account for 86 percent of all fire events. They're also incredibly biodiverse: the Brazilian savanna contains over 12,000 flora and 2,566 fauna, while southern Africa and northern Australia combined have some 5,000 vascular plant species.[3] Traditional practices of burning during the early dry season lowered the risk of larger wildfires later in the dry season. Less dry plant material meant that when lightning arrived at the start of the wet season, there was less chance of a huge conflagration.

Savannas weren't the only landscape in which people used fire, nor were Aboriginal practices of hunting and gathering the exclusive source of burning. Landscape burning was common in various forms of agriculture. Possibly you've heard the most colloquial term: "slash-and-burn," in which forests are cut down, their trunks and leaves and branches burned to ash that fertilizes the ground, while the fire removes most weeds.

But this is only the most generic, ignoring the unique variables at play in every ecosystem—which locals surely did not do. Pyne cites Finnish as an example of one language in which a multitude of terms were used. The vocabulary reflected whether the burning was in forests, if peat turf was added, if the slashing was new, if it was a single-season burning or if it was planned for several years.[4] Yes, fire is unpredictable to a certain extent. But communities who used fire all around the world had a deep knowledge of how it interacted with their local landscape, and they manipulated it to suit their needs.

Then the Enlightenment in western Europe turned open fire from tool to problem and transported those ideas around the world as they subjugated Native populations.[5] When settlers and merchants first began colonizing the coastal Atlantic rainforest of South America in the sixteenth century, they actively used fire to clear land for the production of sugarcane and coffee plantations. But fire suppression became the new standard when fire risked disrupting agriculture and livestock.[6]

This pattern propagated wherever European boots landed. In southern Africa, Europeans extended their grasp into rural landscapes and took control of wildlife and timber, regulating the ability of local people to engage with traditional ecological burning. When the British began colonizing Australia in 1788, the landscape on the southern side of the continent was largely open, with a smattering of trees and shrubs—essentially, a savanna environment.[7] While the settlers used some wide-scale fires to clear land for intensive agriculture, "a fire-fighting mentality came to dominate fire management in Australia, in which fires are seen as a threat to be prevented, or stopped once they start," write researchers Michela Mariani, Michael-Shawn Fletcher, and Simon Connor in an article for *The Conversation*.[8] Cultural burning was banned in many places, undoing a millennia-old relationship between the Aboriginal peoples and their use of fire.[9]

And in the United States, the settler-colonial project of Manifest Destiny and westward expansion required removing Indigenous people from their land as much as possible, whether through direct violence or the violence of unjust legislation. In 1851–52, around the time of the California gold rush, eighteen tribes signed treaties in which they agreed to cede the title to their lands in exchange for payment and reservations set aside for tribal use. But the Senate failed to ratify the treaties, and instead the land was essentially stolen without compensation.[10]

These transparent attempts at Native displacement continued with the Dawes Severalty Act in 1887, which divided tribal lands into individual allotments and resulted in settlers purchasing more than 86 million acres of Indigenous land—all while Native people lost access to traditional fishing, hunting, and gathering sites. This coincided with the National Forest System increasing the territory under its management by 97 million acres. (Today, the US Forest Service manages 193 million acres of land, much of which was once inhabited and stewarded by Indigenous people.)[11]

The USFS instituted a strict fire suppression policy, while also actively working to undermine Native legitimacy and knowledge about the ecological benefits of burning. As part of its campaign to discredit traditional practices, the USFS labeled them "Piute forestry," a racist term conflating the Mono, Southern Paiute, and Northern Paiute, three groups that lived across multiple western states. Visitor maps for Klamath National Forest once asserted that prevention against burning was "not a fire hunt but a man hunt; not fire, but the owner of the hand that lights it, is the public's enemy."[12] USFS fire scientists even sabotaged their own research in order to get the results that showed fire suppression was the only safe strategy to protect timber and natural ecosystems. As environmental sociologist Kirsten Vinyeta writes, "Long before the agency invested any mentionable amount of time or money into researching whether fire has a detrimental impact on forest ecology, they had invested substantial resources into adapting wartime technologies to detect and suppress fires and even experimented with weather modification techniques under Project Skyfire."[13] The agency publicly called fire control "a form of warfare."[14]

The reality was that the land was adapted to fire, that the Indigenous people who had inhabited it since time immemorial had a deep knowledge of the use of fire and other landscape management techniques, and that the ecosystem suffered with fire's expulsion.[15] Science is sometimes held up as an objective set of indisputable Facts, but the production of science is done by humans with biases and agendas. The question that interests me is: Whose voice do we elevate and whose gets silenced? And can we expand our listening ear beyond the normal range of language, into the realm of earth and sky and tree?

Describing the experience of pain is an impossible task, not least because it steals words and replaces them with tears or moans or screams, as Elaine Scarry writes in her famous work on the subject, *The Body in Pain*.[16] It's impossible because no one else can enter into a mutual comprehension of that pain, not with blood tests or ultrasounds or magnets that manipulate the alignment of the body's atoms to map tissues. Even when the source of pain is visible—a swollen joint, a gaping wound—to convey the exact quality of the pain is impossible. But when the pain is largely hidden, a byproduct of the body's inflammatory response working on a scale so small it refuses any meaningful measurements, conveying pain becomes even harder; who will believe it without proof?

I want to write words into my pain because I want you to understand when I say the experience of arthritis unmade me. First, there was its ever-present but ever-changing appearance. One day, swollen fingers on my left hand, preventing me from wearing rings, making it feel as if my flesh and bones were ensconced in a stiff gauntlet. When I intertwined my hands in prayer position, the fingers didn't fit right. There was a stretch at the base, as if I were asking them to make room for some stranger's hand that is too large for my own.

Another day, my fingers were fine, but I awoke to a left elbow so painful and stiff that I could not unbend it. I was Dorothy's Tin Man in need of grease. The sight of my locked elbow scared me so badly that I wrenched it down, which elicited a sparkler of agony up to my shoulder and down to my fingertips. I breathed and bent the joint more slowly, reminding the hinge that it was made for movement.

Most days I struggled to sit in chairs or in the car for any length of time because of the throbbing ache that built up around my hips and lower back. Movement soothed the stiffness, but too much movement triggered even worse pain, as when I visited the Shedd Aquarium with friends and the sharp knife between my hip and ribcage made me cry out and slump against the wall. Sometimes the vertebrae of my upper back burned as if a match had been struck within them. Sometimes a pressurized ache built and built until my spine seemed poised to erupt from my body. There were nights I couldn't sleep because my upper left ribs and sternum throbbed with every breath I took. There were days I could barely walk, because each step was a nail being driven through my heel, or a blade slicing away at the tendons around my kneecaps.

No matter how much Tylenol or ibuprofen or diclofenac I took, something always hurt. It was impossible to believe it would end; it was impossible to know what would hurt next.

For the longest time I tried to cut away the messy, sick parts of myself—the misbehaving thyroid, the upset stomachs, the monthly pains, the arrhythmic heart. When that didn't work I decided to rise above, to be a "super crip" who could compete in triathlons and sword fighting classes despite the broken bits.[17] But I only ended up hurting myself because I still didn't know how to listen when my body screamed "Stop!" I was angry at all the wrongness, the illnesses I didn't think I deserved (as if "deserving" has anything to do with the normal tragedies of living or the systems of oppression we've built).

Arthritis forced me to stop fighting and accept stillness. In unmaking me, it also allowed me to shed beliefs about who I should be, how I should sound, what I should be able to do. The more I sat with my pain, the more I saw that resisting reality would only exacerbate my suffering. I learned to build a deep, powerful love for all the parts of myself. Even an immune system hell-bent on overtaking my organs. Even a nervous system constantly alerting me to the pain of inflammation. My bodymind has been trying to protect me from everything it encounters in the environment, even if that has backfired. In making me sick, it taught me to recognize my needs for care and community and comfort.

Maybe this form of unconditional, spontaneous, wholehearted love that Dr. Martin Luther King Jr. referred to as *agape* always has to start at the individual level. "*Agape* does not begin by discriminating between worthy and unworthy people, or any qualities people possess. It begins with loving others *for their sakes*," King wrote in 1958.[18] He added that *agape* did not mean liking another person or loving their evil deeds; this freed me from the necessity of trying to love any of my diseases (or to stop seeking treatment for them).[19] I do not love feeling pain or undergoing surgery or seeing my body transform—but I've learned to love this body for what it is. For simply being a part of this Earth.

That radical love is how I've been able to write any of my story. It has revolutionized the way I see other humans, the marvelous diversity in our ways of being. And it's made me wonder: How far beyond the borders of my bodymind can I make that love stretch? Can I love the yellow sac spiders living in gauzy pockets at the intersection of ceiling and wall in our house, the kind that once bit me because it was in my sweatshirt, and the bite swelled up and burned? Can I love the ants that infiltrate our laundry room? The doctors who have ignored my symptoms? The storms that make my mouth dry with nervousness? Could I love a river so polluted it caught fire? Can I even love the people running corporations that produce this very pollution, even as I hate their deeds? I'm not sure. But I'm trying.

And if we could all perform such acts of radical love, how might the world change?

I bounced between gastroenterologists for months. All of them told me to see a rheumatologist. But the first rheumatologist was so rude, and so sarcastically dismissed my queries about the possibility of lupus, that I cried for days afterward. The second wanted to start me on fish oil and vitamin D, because none of the inflammatory markers in my blood work were elevated and the MRI of my sacroiliac joint (the conjunction of pelvis and spine) showed no damage. To look at my labs, you wouldn't think anything was wrong. The third rheumatologist also didn't know what was going on but prescribed a disease-modifying antirheumatic called hydroxychloroquine that made my whole body itch. I was trapped in diagnostic purgatory, not sick enough to be taken seriously by a system that requires specific test results, too sick to do anything but sleep and go to doctors' appointments.

What finally made the second rheumatologist decide to put me on an immunosuppressant was that my uveitis kept coming back. This eye inflammation was its own kind of pain, a pickax penetrating to the back of the socket. It came with the risk of blindness.

The doctor still didn't quite know how to diagnose me, but she was willing to guess it was some form of spondyloarthritis, a family of diseases that includes ankylosing spondylitis, non-radiographic axial spondyloarthritis, psoriatic arthritis, reactive arthritis, and enteropathic arthritis.

I received my first dose of Humira at the office. A friendly nurse taught me to press the round tip of the autoinjector against my abdomen and deploy the invisible needle. The medication would dampen my immune system, making me more prone to viruses and infections. Along with other blood tests, I'd now be checked regularly for tuberculosis and hepatitis. Without insurance, the medication would cost more than $7,000 per dose; the doctor gave me an extra dose to take home because of how frequently insurance companies deny the first request to cover the medication. I begged the universe to let this be the medicine that halted the pain. And at first, it did, a little. But then it stopped working, even when my dose was increased. It had been only five months, and I'd already failed my first medication.

The winter I stopped Humira marked a full year of living with symptoms of inflammatory arthritis, and I'd rarely felt so low. The pain was exhausting; I had little energy even when I slept nine or ten hours a night. The COVID-19 pandemic raged on and we didn't yet have a vaccine. I spent the holidays wrapped in a combination of ice packs and heating pads, talking to family through Zoom. I was scared to hope for a better arthritis treatment. I was terrified of getting sicker. For a time, the future ceased to exist.

But my rheumatologist had a plan. She said I'd start a new medicine, an IV infusion called Remicade. I'd come into the clinic and sit with other patients, all of us attached to our IV poles, some sleeping, some reading, some chatting with the infusion nurses. It was the one thing that made me feel the tiniest bit better. My doctor wasn't going to let me face this alone, even if my disease looked different from what she normally saw, even if it didn't fit the standard pattern.

Years later, my rheumatologist, Dr. Erin Arnold, told me that she thinks of her job as needing to prove something *isn't* going on, rather than assuming a patient must be fine because their lab results are within normal ranges and their imaging tests don't show any damage. "Many of the ways that we classically learn about a lot of these presentations is later-stage disease, and frankly, many of the ways that we understood these illnesses in the past are in men," she says.[20] In other words, if she's willing to treat only patients who fit the "classic" disease presentation, a lot of patients will go untreated. Her practice has one of the largest populations of female patients with spondyloarthritis in the country, a category of illness that still isn't well understood in women.[21]

Dr. Arnold built her practice around storytelling and a close patient-doctor relationship. It's the nearest thing to radical love I've ever seen in a doctor's office, this desire to treat each patient as a human—to share book recommendations and holiday cards and provide pain relief when it's requested. I've appreciated it more and more as the years pass, as my disease changes my body, as I fit some patterns and break others. My body holds the illness, but the narrative of disease is shared between us, all the nurses and physician assistants and radiologists and lab technicians and Dr. Arnold herself.

Stoic philosopher Seneca, writing in the first century CE, stated, "We put down mad dogs; we kill the wild, untamed ox; we use the knife on sick sheep to stop their infecting the flock; we destroy abnormal offspring at birth; children, too, if they are born weak or deformed, we drown. Yet this is not the work of anger, but of reason—to separate the sound from the worthless."[22]

The British government passed the Madhouses Act of 1774, which was followed by a series of acts passed throughout the nineteenth century that authorized and funded state-run asylums, places that were essentially prisons.[23] The reasons why a person might be detained in such a place varied widely, writes geneticist

Adam Rutherford in *Control: The Dark History and Troubling Present of Eugenics*. "Asylums housed children and adults, violent criminals, epileptics, people with tuberculosis and other diseases, alcoholics and women with menstrual problems. All these conditions would eventually be the targets of eugenic purification in Germany, America and elsewhere."[24]

At that point, the term "eugenic" hadn't even been invented yet. In 1904, Francis Galton wrote a paper for the scientific publication *Nature* in which he proposed a supposedly scientific way to promote human advancement. "Eugenics is the science which deals with all influences that improve and develop the inborn qualities of a race," he stated. "All would agree that it was better to be healthy than sick, vigorous than weak, well fitted than ill fitted for their part in life.... The aim of eugenics is to represent each class or sect by its best specimens, causing them to contribute more than their proportion to the next generation."[25]

The younger half-cousin of Charles Darwin, Galton had dabbled in a variety of sciences (including meteorology) before reading Darwin's treatise *On the Origin of Species* and becoming obsessed with the idea of selective breeding in humans. He wrote articles and gave lectures on eugenics. At the time of his death, he was working on a novel called *The Eugenic College of Kantsaywhere*, which features a utopian society managed genetically. "Failures had inferior genetic material and were segregated in labor colonies where conditions were not onerous, but celibacy was enforced," wrote geneticist Nicholas Gillham of the novel's imagined society.[26] All members of the society were evaluated for their "fitness," and those with middling scores could procreate with some limitations. The best and brightest were highly encouraged to pair off with one another and produce large families. (If this sounds like something being espoused by modern tech billionaires, that's because it is.)

Galton's ideas found a ready audience in politicians, intellectuals, artists, and public health officials alike. Margaret Sanger, a major proponent of birth control, was a member of the American Eugenics Society and wrote about the need to eliminate the disabled, the criminals, and those of lower intelligence from society, since they were "human weeds which threaten the blooming of the finest flowers of American civilization."[27] Multiple white supremacists in the US invoked eugenics to support their fears of a "great replacement," from T. Lothrop Stoddard's *The Rising Tide of Color Against White World-Supremacy* to Madison Grant's *The Passing of the Great Race*. British author D. H. Lawrence made perhaps the bluntest connection between eugenics and genocide, dressed in flowery language:

> If I had my way, I would build a lethal chamber as big as the Crystal Palace, with a military band playing softly, and a Cinematograph working brightly; then I'd go out into the back streets and the main streets and bring them all in, the sick, the halt and the maimed: I would lead them gently, and they would smile me a weary thanks, and the brass band would softly bubble out the "Hallelujah Chorus."[28]

The mind jumps to the concentration camps of Nazi Germany, but history demands a stopover in the United States first, where the Eugenics Record Office opened in 1910 with the dual goals of reducing immigration and using forced sterilization for "defectives," while also compiling national records of families and their health. Office superintendent Harry Laughlin published *Eugenical Sterilization in the United States* in 1922 to codify whose reproductive ability should be eradicated.[29]

In 1927, the Supreme Court heard the case of *Buck v. Bell* and upheld Virginia's Eugenical Sterilization Act. Carrie Buck, who stated that she was raped and impregnated by her foster parents' nephew, was deemed "feeble minded" and ordered to be sterilized.[30] In his opinion for the case, Justice Oliver Wendell Holmes wrote, "It is better for all the world if, instead of waiting to execute degenerate offspring for crime or to let them starve for their imbecility, society can prevent those who are manifestly unfit from continuing their kind."[31] By 1931, more than twelve thousand men and women were forcibly sterilized in thirty states; at the end of the twentieth century more than seventy thousand people had been sterilized, the bulk of them women of color.[32]

In 1933, Laughlin's model was co-opted by Hitler's government. Under the Nazis, eugenics was taken to its extreme ends, with the Third Reich murdering some six million Jews as well as millions more ethnic Poles, Romani, people with disabilities, German political dissidents, and Soviet prisoners of war.[33]

Who was in control and who was being controlled depends entirely on which voices are considered reliable. Who is more trustworthy, the judge or the criminal? What if the criminal is a young girl who has been raped and locked in an institution against her will? What if the judge has been shaped by decades of eugenic discourse in popular and intellectual culture, and he wants to ensure the continuation of an ethnonationalist America? The trustworthiness we assign to every individual's narration of their experience is influenced by dozens of factors: their age and race and education and history, their psychology and spirituality and personality. Perhaps most important is the context in which they try to assert themselves. Caught in a lie, will a child double down, invent something new, or confess? Caught in a political system like eugenics, will a patient (victim) maintain their story, temper it to appease their adjudicator, or lie to save themselves?

Suppress the fire to make the forest beautiful. Suppress the disabled to make society wholesome. Prescribe morals to everything—cultural burning, marriage, reproduction—and let society shackle itself. As Rutherford writes, "The persistent theme of the fall of existing civilizations is a lodestone to which the eugenicists are all magnetically drawn."[34] I write these words just days after Donald Trump, still a presidential candidate, held a rally at which the speakers reinforced fears of the white majority being replaced by immigrants. Weeks later, he'll be elected again despite the intolerance of his eugenicist message—perhaps because it's a message that resonates with so many.

Nazi gas chambers and medical experimentation may have scared some countries off the more obvious form of eugenics, but even that wasn't enough to end it. Early in the COVID pandemic, the advent of vaccines marked a downturn in the number of deaths—but those vaccines did not completely prevent death or stop the spread of disease. No less a personage than Rochelle Walensky, the CDC director at the time, stated in early 2022, "The overwhelming number of deaths, over 75 percent, occurred in people who had at least four comorbidities. So, really, these were people who were unwell to begin with."[35] While Republican and right-wing figures took Walensky's statement as evidence the government was lying about who had been dying of COVID since the start of the pandemic (creating yet more misinformation), disability advocates and organizations saw her words as a sign that our lives don't matter—that it is encouraging news to learn only those who are "unwell to begin with" are still dying.[36]

One evening, sitting indoors with my pain-ridden body, trying to avoid thinking about COVID deaths and when the virus might come for me, I was flipping through books as a distraction. I reopened a yellow-paged manual from my mother-in-law's childhood, the *Book of the Camp Fire Girls*. When she'd first showed me the book, it reminded me of my participation in the Brownies—going to summer camp, learning to build fires, earning badges. The Camp Fire Girls seemed like an earlier iteration; the book was full of drawings of cartoon girls in skirts and bandanas.

There was the prerequisite cultural appropriation—full pages dedicated to picking one's "Indian Name" and drawing "American Indian symbols"—as well as more useful information about tree identification and building fires. When I flipped back to the copyright page to learn the year of publication (1958), I realized I'd missed "The Law of the Camp Fire Girls," which listed a number of principles, from "Worship God" to "Seek Beauty." Near the bottom of the list came "Hold on to Health," which seemed a strange thing to include alongside directives like "Pursue Knowledge" and "Be Trustworthy."

Made curious (or perhaps suspicious) by the emphasis on health, I began doing more research. The organization still exists under the shortened name of "Camp Fire," and its history page includes extensive information about the places allotted to disabled campers early on, including program materials translated into Braille in 1914. By 1952, the organization's national office even released a booklet called "Services with and for Handicapped Children."[37]

But an article from the *Journal of Heredity,* written in 1915 by A. E. Hamilton, a member of the Eugenics Record Office, singles out the Camp Fire Girls as a prime example of the power of eugenics in action. The founders of the organization, Dr. Luther Halsey Gulick and Charlotte Vetter Gulick, are quoted as saying, "The ranks should be recruited from those who have ability to do and to help rather than from those who need help."[38] While Hamilton notes that the word "eugenics" appears nowhere in the organization's materials, he finds plenty of evidence to suggest participation in the group "means something for the future of the race," implying that the thousands of girls who may come to participate will be more healthful and able mothers who will help perpetuate whiteness and even learn to vote against certain types of immigration, such as "the Jukes and Ishmaelites and gunmen who flood into our country through the leaks in our immigration laws."[39]

The highest attainable rank for members of Camp Fire at the time was "Torch Bearer." A description of their role states, "That light which has been given to me, I desire to pass undimmed to others."[40] Nothing about this strikes me as malicious; an understanding of fire and its undimmed light was the gift of Prometheus to humankind. Where Hamilton sees fire as a metaphor for passing on only the best genes, I interpret it as a lesson about the strength of community. We can't keep light for ourselves alone; it grows with the sharing. Given how much Camp Fire seems to have been committed to inclusivity from relatively early in its inception, I want to blame Hamilton for co-opting the group to serve his own interests. But maybe eugenics was an unconscious element of the group's foundations. It's hard to avoid imbibing cultural beliefs when they're all around, in the very air you breathe. I even understand the desire to "Hold on to health" because it seems true that no one *wants* to get sick.

The trouble is that so much of illness and disability is beyond our control. When our environment is filled with pollutants and we're exposed to viruses (like COVID) that trigger autoimmunity and we live with the stress of social and financial precarity, we can't simply stay healthy through sheer force of will and manifestation. Chronic illness is affecting ever greater numbers of people of all ages and doing so as climate change is making it more likely that even *more* people will get sick and become disabled.[41] We can't focus our energy and attention solely on how to maintain health. What deserves equal attention is choosing to build a world dedicated to supporting the currently sick and disabled.

OUR DISABLED ECOLOGIES

We borrowed from Latin for the term "inflammation": to set on fire. And it was Roman encyclopedist Aulus Cornelius Celsus who first listed the four signs of inflammation as swelling, fever, pain, and redness (ignoring or omitting skin variations in humans and that inflammation presents differently in dark skin).[1] But we have evolution to thank for this process that keeps us alive. Infection, illness, and injury all trigger the body's immune system to respond, creating inflammation. While fever, pain, and swelling are unpleasant to experience, the end goal is to return the body and its tissues to homeostasis. As Scottish surgeon John Hunter astutely wrote in the eighteenth century, "Inflammation is itself not to be considered as a disease, but as a salutary operation, consequent either to some violence or some disease."[2] Without these little fires produced by our body's immune response, we would die. Think of them as the human equivalent of small-scale landscape burning; savanna ecosystems also suffer without fire.

But like wildfires, inflammation in the body can grow to harmful levels. Sometimes a bacterial or viral pneumonia will result in sepsis, a life-threatening hyperinflammatory response in which the immune system itself causes organ damage and sometimes organ failure while at the same time, other elements of the immune system are underactive, leading to immunosuppression and the risk of contracting a secondary infection.

Another form of immune dysregulation is seen in autoimmune diseases, in which the immune system targets the body with inflammatory attacks. There are over one hundred known autoimmune diseases, and they can damage virtually every part of the body. This state of chronic inflammation can be initiated by infections, air pollution, stress, pesticides and other toxins, and even certain foods. And while diagnosing specific autoimmune diseases is challenging, markers of autoimmunity are increasing.[3] So, too, is the annual incidence of autoimmune diseases, at a rate ranging from 3 to 7 percent globally.[4] The trouble is there are no national registries of autoimmune diseases, no firm definitions of the criteria for many of them, and we don't understand the immune system well enough to comprehend the mechanics behind why it's going wrong.

"The more we learn about immunology, the more we realize it actually doesn't follow the rules we thought," says Yasmin Mohseni, a scientist at a biotech company who did her PhD work in immunology and immunotherapy.[5] While scientists understand big-picture things about the immune system, such as the role of different organs in producing immune cells and which cells are more specialized, questions remain when it comes to how immune cells transition between different forms and why the immune system goes haywire in autoimmune disease. Mohseni, who posts regularly on Instagram to debunk wellness claims about "immune-boosting supplements" and how to "fix your gut and fix your disease," is keen to explain just how rapidly our knowledge of the immune system is changing. At one point she started revising an immunology textbook that she loved because so many things in it were becoming out of date. When others asked her whether they should get a copy of the book, she had to tell them no, because so much of the information was wrong.

This recognition of our limited knowledge about the immune system is shared by Frederick W. Miller, a scientist emeritus with the National Institutes of Health and former head of the Environmental Autoimmunity Group. "Half of what we think we know is wrong, we just don't know which half," he says.[6] Miller spent his career investigating sources of autoimmune disease and the nexus between genes and the environment. He also gave the first talk on autoimmunity and climate change at the American College of Rheumatology and has written extensively about the many ways that climate change is contributing to the development of new disease and the exacerbation of existing disease symptoms.

"It's been an uphill climb all the way, and there's still a lot of resistance to these things," says Miller. But he adds that he's sought out difficult projects and challenges, in part because they're interesting. "I knew I would only be probably chipping away at the iceberg in my short time, but I decided to do it and tackle it because I appreciated at the time that the environment was the major risk factor for the development of autoimmune diseases."

Hearing Miller talk about his lifelong work makes me think of the Latin origins of "radical." The word, which comes from *radix*, means "root." In more figurative language, it refers to the essential or fundamental qualities of something—not a particular political orientation. Miller's work is radical because it seeks to understand the origins of autoimmune disease, whether that be environmental exposures, infectious diseases, pollutants, or lifestyle factors. He is going deep as well as broad, prying apart myriad factors to see which of them most readily feeds the strangling vine of autoimmunity.

Political activist Angela Davis used such language to describe the source of oppression. "If we are not afraid to adopt a revolutionary stance—if indeed we wish to be radical in our quest for change—then we must get to the root of our oppression," she wrote. "After all, 'radical' simply means grasping things at the root."[7]

There are many things about my own experience of illness for which I'll never be able to locate a specific root cause (and I've learned to be wary of wellness influencers who use that language). There are too many unknowns. And because no one has uncovered the explanation for autoimmune diseases, whether I get sicker is also beyond the scope of my understanding or control.

What is obvious is that our planet is in a bad state (and I'm culpable simply by being an upper-middle-class American who consumes more than my share), that the effects of climate change are uneven, but we're all in for a hard time. Acknowledging all the history that led us here is heavy, hard to bear. But I want to bring all my emotions to play in the issues of illness and the environment. Accepting my denial, my fear, my rage, and my grief allows me to arrive at radical love. And that love of life, *all life*, offers clarity. Change is necessary. Change is already happening; we must continue to choose change and to envision a better future. As artist and activist Sunaura Taylor writes, "In this Age of Disability we can fear disability and see it as the antithesis of the future we wish to build, or we can center the damaged, the vulnerable, the ill and the injured, the contaminated and the poisoned, and build a new world from there."[8]

The Age of Fire is here, whether or not we've accepted it. Forest fires have been burning an increasing amount of land since 2001, to the point that they now consume twice as much tree cover as they did twenty years ago.[9] Climate change is creating patterns of heat and drought that dry out foliage, transforming them into crisp kindling. This expanded burning is a global phenomenon: an increase of 22 percent in Northern Australia, 15 percent in western North America, 29 percent in southeastern South America.[10] Wildfire deaths are mounting in the US, and the smoke from fires worldwide is contributing to around half a million deaths a year.[11] The fires aren't just getting bigger: they're getting faster. In August 2023, a single fire caused by broken power lines escaped containment and burned the town of Lahaina on Maui in just two hours, destroying more than two thousand structures and killing 102 people.[12] As one study notes, "Some of the most deadly and destructive wildfires in US history have occurred in recent years, with most having the common characteristic of extremely rapid growth. . . . Moreover, the frequency of fast-growing fires is predicted to increase by ~50 to 200% with projected warming."[13]

Yet there are some glimmers of hope. Indigenous communities are receiving funding, government support, and media attention to expand and revitalize cultural burning practices. In Australia's Northern Territory, home to the largest intact tropical savanna ecosystem on Earth, dozens of Aboriginal-owned and -operated savanna fire projects manage the land by lighting "cool" fires early in the dry season.[14] This traditional technique has reduced the frequency of dangerous, late dry-season wildfires.[15] While prescribed burning as practiced by government organizations is aimed at hazard reduction, proponents of cultural burning point to it as "a holistic approach at landscape management . . . in which the health of people is linked to the health of the world around them."[16]

"The fire has been a part of our lives since the beginning of time. We probably invented fire," says Robin Dann, a Wunggurr Ranger, in a video released by the Australian Department of Climate Change, Energy, the Environment and Water.[17] The success of these cultural burns in protecting landscapes and wildlife has allowed groups to become involved in the carbon credit industry, earning credits by preserving the landscape so more carbon isn't released in larger late-season savanna fires. The funds provide more jobs and opportunities for training in cultural burning.

In the United States, tribes in California and Oregon are renewing traditional practices of cultural burning, inviting students and government agencies to join them.[18] In central California, the Amah Mutsun Tribal Band employs cultural burning to restore the coastal prairie grasslands.[19] The North Fork Mono Tribe is demonstrating how cultural burning can rejuvenate plants like the three-leaf sumac.[20] And further north, in the Klamath Basin, the Karuk and Yurok Tribes are expanding beyond Western science to incorporate traditional ecological knowledge into landscape management and cultural burning.[21]

Melinda M. Adams, a member of the N'dee San Carlos Apache Tribe and environmental scientist, has written extensively about the ways that Indigenous knowledge improves ecosystem health. She has led or participated in dozens of cultural burns and sees ceremonial fires as a way for "young people [to] brace for the onset of megafires now prevalent throughout California."[22] Adams considers the work of Indigenous women in particular as crucial for healing native plants and soil, and creating connections between humans, animals, and the environment.[23]

Collaborating with Adams is interdisciplinary scholar Erica Tom, who describes the trauma of seeing her childhood home destroyed in 2017 wildfires: "I walked through the place where my family room had been, where my bed had fallen with the second story: the iron frame had been twisted by the flames, as if gripped by a giant's hands. The refuge of those walls, where I could walk through the door and return to the safety of *home*, was destroyed."[24] She dealt with symptoms of post-traumatic stress in the aftermath of the fire. It was only in joining tribal communities to learn the practice of cultural burning that she began to once more feel comfortable in the fire-prone landscape.

Both women refer to the work of Australian philosopher Glenn Albrecht, who created the term "solastalgia" to describe the profound sadness of our current relationship to the earth. In coining the word, which initially arose from thinking about "nostalgia," Albrecht sought to convey a precise form of melancholia. "Solastalgia is the pain or sickness caused by the loss or lack of solace and the sense of isolation connected to the present state of one's home and territory," Albrecht writes.[25] "The defeat of solastalgia and non-sustainability will require that all of our emotional, intellectual and practical efforts be redirected towards healing the rift that has occurred between ecosystem and human health, both broadly defined."[26]

When I first began researching cultural fire and burning practices that emphasized restoration, I hoped to make a pilgrimage of sorts to Northern Australia and volunteer with one of the Aboriginal groups there. But my body was experiencing one of its periodic flares, a wildfire that needed time and medication to calm. I was too ill for any travel for months and missed the early dry season. I considered the possibility of traveling instead to California, a much shorter voyage, but even that proved impossible for a time. It occurred to me only after accepting this inability to venture out that what I was seeking elsewhere was a cure to my own solastalgia. I wanted to meet people forming this intimate connection to their home environment so that I could bottle some of that empowerment and bring it back with me. I wanted to stop feeling such desolation.

I began reading about fire regimes around the Great Lakes and was surprised to learn that here, too, fire has a long history. Indigenous people used fire all around the region; the prairies and open woodlands were maintained by fire and may have been burned annually.[27] Indigenous people in Michigan, Minnesota, and Wisconsin are all working to bring fire back into the varied ecosystems around the lakes. Anishinaabe people across this region traditionally used fire to assist with the growth of blueberries and increase the resilience of the ecosystem. The project in Wisconsin, Nimaawanji'idimin Giiwitaashkodeng (the name translates to "We are all gathering around the fire"), aims to uncover the ways in which Native people used fire for cultural and ecological purposes.[28]

Even in Chicago, fire is part of the landscape. The Forest Preserves of Cook County oversee prescribed burns in different parks to combat invasive species like honeysuckle and buckthorn and improve the growth of wildflowers and grasses. Although the Illinois Prescribed Fire Council reports that a much greater amount of burning is needed to maintain healthy ecosystems across Illinois, the places that have been burned for decades show remarkable biodiversity.[29] One hundred miles west of Chicago, the Nachusa Grasslands is a 4,100-acre nature preserve where prescribed burns have been used for more than three decades.[30] In Illinois, only 1 percent of historical tallgrass prairie still exists; at one point, nearly one-third of North America was prairie, and today this ecosystem is endangered. But at Nachusa, conservation managers have brought back bison, and research is showing that the use of fire even has a positive impact on invertebrates like moths and bees.[31]

Fire destroys and it creates. How we harness it now, whether we choose to learn from Indigenous tradition or not, and how we change our relationship with lithic fire will reshape the entire planet. It is already doing so, with terrible consequences. But we still have time to change.

"We need to get right with fire," says Stephen Pyne, the pyromantic who was puzzled by the need to extinguish lightning-ignited fires in a landscape made for them when he worked on a fire crew. "Yeah, there are some bad fires, and climate change is making it worse. But we're going to have to reconnect with our fire heritage if we want to be able to live in a habitable world."

November has been unusually warm, like every other month in 2024, until a sudden cold snap hits. Snow falls, enough to stick to grass and rooftops and survive the day. The oak trees in our yard release the last of their leaves. I switch to my winter coat and begin wearing leggings under my jeans. We huddle indoors against the darkness that falls before 5:00 p.m. This is the month a friend and I have chosen for our fire ritual.

The idea comes from an earlier conversation, in which we both lamented the lack of rituals available to those going through chronic illness and disability. We celebrate the anniversary of our birth but have nothing for the anniversary of a diagnosis. There are coming-of-age events, no coming-of-illness. What do we do with all the feelings that arise after a bad doctor's appointment, or before a big surgery, or with the loss of an ability, or when we finally find the right medication or get approved for disability accommodations? We both feel the lack in our lives, so we decide to create something. We'll write a zine full of little rituals, and we'll aim to host four annually, one for each season. Because she lives in Seattle and I'm in Chicago, they'll be hybrid events, virtual and in-person. She'll light a fire for all the virtual attendees, while I'll build one for people that can come in-person. We call the first ritual "Burn it Down" for the fire that's meant to consume our many griefs.

The night of the event, I bake pumpkin chocolate chip cookies and build the bonfire in our fire pit. Logs stacked in a teepee formation, the dead dry heads of hydrangeas serving as kindling, along with twigs that fell from the bare-branched trees in a recent windstorm. I use a lighter, then resort to the kitchen blow torch that we usually employ for scorching the sugar on top of halved grapefruits. It feels like cheating, but I'm too nervous to fuss with matches. I want the event to feel meaningful to the people attending, and we need a good fire.

The logs blaze, heat and smoke rippling outward from the flames, providing a slight blanket against the cold darkness. People begin arriving in person and on Zoom. Despite the clouds, there's no snow or rain, and when we officially open the ritual by offering handfuls of leaves to our respective fires, they burn with a voracious crackling.

The ritual is simple: everyone writes their griefs on paper and then burns them, some in the two main fires, others with candles at home. I choose only one grief to share, a new diagnosis, but there are so many others weighing down my heart. In terms of health, 2024 has been one of the hardest years I've ever experienced. Four new diagnoses, several with treatments that require needles penetrating unpleasant places. Thirty weeks of physical therapy, first to help with recovery from hip surgery, then for my arthritis symptoms. I've seen twenty different doctors—and I don't mean just twenty appointments, I mean twenty individuals whom I saw for close to one hundred appointments. I've had X-rays and ultrasounds and one MRI; one procedure under general anesthesia, another under twilight sleep. I'm so very tired from carrying the weight of it all, from fighting insurance and scheduling appointments and listening to unkind doctors dissect my body and gritting my teeth through painful treatments and paying thousands of dollars to keep the bigger pains at bay.

I could speak for hours, venting it all, but I limit myself to the few minutes we've given everyone. It's only when I burn the paper holding my single grief that I imagine all the rest of them pouring down. Maybe the fire feels that energy coming off me, because the crumpled paper shoots up into foot-high flames, larger than you'd expect from something so small.

The ritual lasts an hour and a half. There are tears and heart emojis and the warmth of having your grief seen and swallowed by flames. At the end, after we say our goodbyes and I warm my cold fingers indoors, grateful contentment fills my limbs. The grief isn't gone, but it is lighter. I feel lifted by a community of people who understand suffering and know how important it is for that pain to be seen. We are witnesses for one another, offering radical love to all difficult bodies. For weeks afterward, my coat and hat and scarf all smell like bonfire smoke, and I inhale the scent and remember what we all burned. I think of fire as the center of that community gathering, welcome and necessary.

It reminds me of Glenn Albrecht's more recent neologism, "soliphilia," which is "the love of the totality of our place relationships, and a willingness to accept the political responsibility for protecting and conserving them at all scales."[32] I find community in my environment when I plant native wildflowers and shrubs in the yard, when I pause in my walk to watch a Cooper's hawk eating another bird and casting its feathers down from the sky, when I make it a mission to rescue the periodical cicadas ambling across the sidewalk and place them on trees where they won't be squashed. Being chronically ill means I can never do as much as I want to fight for this ecosystem, that my tests and treatments will generate more plastic waste, that I need to drive on a fairly regular basis to reach my appointments.

But illness has also taught me determination and creativity and resilience and the importance of community. In learning to love my body even when it's hurting, and sick, and broken, I've learned to love an overheated, polluted, damaged planet. I believe that no person is dispensable because of their disability, just as no environment deserves to be buried or forgotten because of what's been done to it. There was a time when Lake Erie was thought to be "dead" due to pollution and eutrophication, but today it is full of life. I love the lake even with its algal blooms and non-native species and anoxic zones. I want to protect it from further damage just as I want to protect the people and other creatures living within and around it. I want to take inspiration from Sunaura Taylor's work and words, to believe that "while disability is evidence of crisis, it is also evidence of connection," that it is "the consequence of injustice, but also a propeller of justice."[33]

I want to say that I am valuable not in spite of my illnesses but because of them. I want the world to recognize all the wisdom, multiplicity, and power of the disabled and chronically ill experience. I want you to know that I have a good life even if I also have bad days, bad months, hard years. To live in a sick body is not easy, ever. I excavate pockets of comfort, I rest my head on the shoulders and arms of my friends, I make dark jokes on the worst days. ("We are in many many troubles for the moment, so many that grief loses its dignity and bursts out laughing," wrote poet Robert Frost in a 1931 letter to his friend, which is exactly how I felt when I broke my big toe a few weeks after coming off crutches for hip surgery—how I feel every time something else gets added to the list of incurable illnesses my body develops.)[34]

I value what I've learned from illness and I still wish for cure. I contradict myself. I want disability justice, the end of private health insurance, care for whoever needs it whatever their condition, care uncoupled from capitalism. I want all these things because I believe the world will be better when we center the needs of the most vulnerable, but this is also a desire for myself.

Radical love is an emotion, but it is also an orientation toward collective liberation. We must understand the roots of these exploitative, destructive systems of power so that we can dig them up and leave behind no harmful fragments. But the harder work, I think, is envisioning what to grow in their place. It requires pushing our imaginations to expand beyond what we know of the world, beyond our own experiences. We have to choose, over and over again, to love our planet so deeply that we stop elevating individuals above all else. We are human; we will all experience anger and fear and grief and loss. We will all die. Our love has to exist within this ecosystem instead of fighting it.

When I let the force of my love direct my imagination, I see a world without cages, cities filled with collaboration between humans and flora and fauna, medical research led by patient-doctor partnerships. I see fires coaxed to life across the landscape for ecosystem restoration and ceremonial connection. I no longer smell diesel or gas exhaust from vehicles, or the fumes of coal plants and oil refineries. I know the names of birds and trees and insects of my biome, and I understand their seasonal patterns. My use of technology no longer relies on extraction but is created through symbiosis. There are many sick and disabled in this world, but we are not ostracized. We are cared for as valuable members of the community.

I do not have the expertise or knowledge to say how we get to this version of the world, but I believe human ingenuity could be put to better use if it were led by radical love. The opening months of the COVID pandemic showed us an alternative: masking for one another, increased Medicaid coverage, a temporary national moratorium on evictions. Why didn't we use that momentum to push for a greater overhaul of the healthcare system and expanded social safety nets? Why, when we saw the reduction in air pollution with reduced travel during lockdowns, did we not push harder for more environmental protection?[35] To be human is to be vulnerable, forever balanced on the edge between sickness and health, and that human vulnerability is paired with the health of our ecosystems.

We don't have to choose nihilism in the face of certain death, any more than we must choose inaction in the face of an ongoing climate crisis. We can choose to change. This feverish world is not the only option. The lessons I have learned from illness are easily applied to the environment. As Taylor writes, "Processes of ecosystem impairment, illness, and mutation are not mere metaphors, but rather could generatively be understood as part of the disability experience: *ecological disablement*."[36]

It's possible to build a smaller but richer life, to spend less energy in the frantic scramble for more, to embrace stillness. It is possible to live with pain. It is possible to survive heartrending grief. It is possible to let go of dreams without the promise of bigger ones. It is possible to build connections with humans and the more-than-human world, to find ways of living that support us all. It is possible to harness rage to demand a better world. It is possible to care with your whole heart and whole mind and whole body, no matter what weather system you find yourself in.

ACKNOWLEDGMENTS

Writing a book, like living with chronic illnesses, is something one must do alone. But in the best cases, you're surrounded by a community that offers collaboration and assistance along the way. I've been fortunate to have many people supporting the book-writing process and keeping my body functional enough to do this work.

I didn't think I'd ever write something that included so many personal anecdotes, and I wouldn't have had the confidence to do so without the encouragement of many mentors and editors. Enormous thanks to Esmé Weijun Wang for creating the Unexpected Shape Academy. Her classes and guidance made me believe I could tackle the difficult stories from my own life. Sarah McColl's class on experimental forms in personal narrative illuminated a pathway for the memoiristic elements of my book. Tajja Isen took some of my earliest work about chronic illness for *Catapult Magazine* and not only gave valuable edits but made me feel safe enough to share vulnerable moments with readers. Daniel Gross did the same at the *New Yorker* digital, helping me write an article that directly led to this book.

Thank you to Jennifer Carlson for encouraging me to undertake this project and finding a home for it, and to Joanna Green for offering insightful feedback while also making sure I was taking care of myself. Special shoutouts to Alison Rodriguez for wrangling paperwork and blurbs, and Kyrianna Bolles for the beautiful portrait on the cover. The whole team at Beacon Press have been incredible—the designers, the sales and marketing team, and the copyeditors. Thank you for bringing my book into the world.

I'm so grateful to Columbia Journalism School, Harvard University, and all the staff and judges who administer the Lukas Prize Project. Without the financial support of the Book-in-Progress Award, I'm not sure I'd have been able to travel for research or hire a fact-checker. Speaking of whom, Wudan Yan provided her incomparable talents in fact-checking a book that covers many areas of scientific and historical research. Any remaining mistakes are my own.

The Logan Science Journalism Program at the Marine Biological Laboratory gave me incredible hands-on experience in gene editing and biomedical science. I'm so grateful to the staff and all the scientists who taught us and shared their work. I only wish I could've fit more of those stories into the book! I had the most wonderful experience with all the other participating journalists too: a group of smart, funny, and kind people, whose work I hope to keep reading for many years to come.

My residency at Yaddo gave me the time and space to write some of the final parts of the book alongside an incredible group of artists. Having the opportunity to share an early version of my work with other writers, poets, artists, filmmakers, actors, musicians, directors, and choreographers was a gift. Special thanks to the entire kitchen staff, who made sure I had delicious gluten-free, dairy-free food to eat.

I couldn't have written this book without the knowledge of the many scientists and historians who spoke with me. Thank you to everyone who agreed to calls, and especially to all the researchers working at Devils Hole with *Cyprinodon diabolis*. Your passion for this quirky fish makes me believe humans can have a good relationship with the creatures of the world.

I'm indebted to the work of other disabled and chronically ill writers. Many of them appear as sources, but I'd additionally like to thank Alice Wong, Riva Lehrer, Jen Deerinwater, Porochista

Khakpour, Leah Lakshmi Piepzna-Samarasinha, and Suleika Jaouad. I also want to highlight the *Sick Times* for their important coverage of Long COVID. Your work has inspired me, taught me, challenged me, and provided an ecosystem in which to explore my own experiences with illness and disability.

So many, many friends offered comfort, wisdom, and support throughout this process. Thanks to Jackie Mansky and Rachel Gross for the voice messages and Zoom brunches. To Sarah Mercer for letting me cry on your shoulder when it feels too hard. Thanks to Maya Wei-Haas for all the dog pics and venting and celebration. Abby Geni is the best sick writer friend I could ask for. Jessica Leigh Hester helped me track down papers and commiserated about SI joints being the worst pain in the butt. Anna Cechony also helped with paywalled sources and is the most joyful collaborator in creating rituals for chronic illness. Jaime McConachie and Krista Rand have shared texts and stickers and emotional support. All of my Thrive friends are the best thing that has ever come out of the Internet; thank you for the advice, the cards, the dance parties, the crafting, the love.

Dozens of doctors, nurses, PTs, and OTs have helped me manage my body over the last fifteen years. Special thanks to Dr. Tu, Dr. Arnold, Dr. Brady, Dr. Robin, Dr. Dori and Dr. Milhouse (as well as all the PAs, nurses, lab techs, and staff at your offices!) for providing compassionate care.

Thank you to the Great Lakes ecosystem for being home most of my life. Thank you to all the Indigenous people who stewarded this environment for millennia, and to all the people still working tirelessly as caretakers.

Biggest thanks to my family, who have taken care of me after surgeries when I'm feeling the worst. To my cat, Hibou, for being the best model of how to prioritize rest when your body is falling apart. And to Kevin, again, for driving me to appointments, fetching heating pads and ice packs, checking every menu to make sure I can safely eat, and making disabled life much more manageable. I love you.

NOTES

PROLOGUE: WEATHER WALKS

1. Margaret Price, "The Bodymind Problem and the Possibilities of Pain," *Hypatia* 30, no. 1 (Winter 2015): 268–84, https://www.jstor.org/stable/24542071.
2. Polly Atkin, "Threatening Rain: On Bodies, Bad Weather, and Bad Clothing," in *Moving Mountains: Writing Nature Through Illness and Disability*, ed. Louise Kenward (London: Footnote Press, 2023), 108.
3. Susan Sontag, "Disease as Political Metaphor," *New York Review of Books*, Feb. 23, 1978, https://www.nybooks.com/articles/1978/02/23/disease-as-political-metaphor/.
4. D. Gayle DeBord et al., "Use of the 'Exposome' in the Practice of Epidemiology: A Primer on Omic Technologies," *American Journal of Epidemiology* 184, no. 4 (Aug. 2016): 302–14, https://doi.org/10.1093/aje/kwv325.
5. Sarah Manguso, *The Two Kinds of Decay: A Memoir* (New York: Farrar, Straus & Giroux, 2008), 183.
6. Manguso, *The Two Kinds of Decay*, 183.
7. John Donne, "No Man Is an Island," All Poetry, https://allpoetry.com/No-man-is-an-island.

WHEN THE TEMPERATURE HAS TEETH

1. Tamás Dávid-Barrett and Robin I. M. Dunbar, "Bipedality and Hair Loss in Human Evolution Revisited: The Impact of Altitude and Activity Scheduling," *Journal of Human Evolution* 94 (May 2016): 72–82, https://doi.org/10.1016/j.jhevol.2016.02.006.
2. Daniel Aldea et al., "Repeated Mutation of a Developmental Enhancer Contributed to Human Thermoregulatory Evolution," *Proceedings of the National Academy of Sciences* 118, no. 16 (Apr. 2021): e2021722118, https://doi.org/10.1073/pnas.2021722118.
3. Kazuhiro Nakamura and Shaun F. Morrison, "Central Efferent Pathways for Cold-Defensive and Febrile Shivering," *Journal of Physiology* 589, no. 14 (July 2011): 3641–58, https://doi.org/10.1113/jphysiol.2011.210047.
4. Jennifer Vanos et al., "A Physiological Approach for Assessing Human Survivability and Liveability to Heat in a Changing Climate," *Nature Communications* 14, no. 1 (Nov. 2023): 7653, https://doi.org/10.1038/s41467-023-43121-5.
5. Thomas S. DaVee and Edward J. Reineberg, "Extreme Hypothermia and Ventricular Fibrillation," *Annals of Emergency Medicine* 9, no. 2 (Feb. 1980): 100–102, https://doi.org/10.1016/S0196-0644(80)80339-8.

6. Richard Shine, "Reptiles," *Current Biology* 23, no. 6 (Mar. 2013): R227–31, https://doi.org/10.1016/j.cub.2013.02.024.
7. Bayard H. Brattstrom, "Body Temperatures of Reptiles," *American Midland Naturalist* 73, no. 2 (1965): 376–422, https://doi.org/10.2307/2423461; Ary A. Hoffmann, Steven L. Chown, and Susana Clusella-Trullas, "Upper Thermal Limits in Terrestrial Ectotherms: How Constrained Are They?" *Functional Ecology* 27, no. 4 (2013): 934–49, https://doi.org/10.1111/j.1365-2435.2012.02036.x.
8. Yuuki Horii, Takahiko Shiina, and Yasutake Shimizu, "The Mechanism Enabling Hibernation in Mammals," in *Survival Strategies in Extreme Cold and Desiccation: Adaptation Mechanisms and Their Applications*, ed. Mari Iwaya-Inoue, Minoru Sakurai, and Matsuo Uemura (Singapore: Springer, 2018), 45–60, https://doi.org/10.1007/978-981-13-1244-1_3; Øivind Tøien et al., "Hibernation in Black Bears: Independence of Metabolic Suppression from Body Temperature," *Science* 331, no. 6019 (Feb. 2011): 906–9, https://doi.org/10.1126/science.1199435; Rochelle Buffenstein et al., "The Naked Truth: A Comprehensive Clarification and Classification of Current 'Myths' in Naked Mole-Rat Biology," *Biological Reviews* 97, no. 1 (2022): 115–40, https://doi.org/10.1111/brv.12791.
9. Arjan Mann, Bryan Gee, and Jason D. Pardo, "A Fossil Discovery Reveals the Earliest Relative of Modern Mammals," *The Conversation*, May 26, 2020, http://theconversation.com/a-fossil-discovery-reveals-the-earliest-relative-of-modern-mammals-138103.
10. Anne Mette Frøbert et al., "Hypothyroidism in Hibernating Brown Bears," *Thyroid Research* 16, no. 1 (Feb. 2023): 3, https://doi.org/10.1186/s13044-022-00144-2.
11. Caitlin P. Wells et al., "Life History Consequences of Climate Change in Hibernating Mammals: A Review," *Ecography* 2022, no. 6 (2022): e06056, https://doi.org/10.1111/ecog.06056.
12. "Ancient Myths Inspired by Fossils," *Biodiversity Heritage Library* (blog), Oct. 13, 2015, https://blog.biodiversitylibrary.org/2015/10/ancient-myths-inspired-by-fossils.html.
13. "Earth Can Regulate Its Own Temperature over Millennia, New Study Finds," *MIT News*, Massachusetts Institute of Technology, Nov. 16, 2022, https://news.mit.edu/2022/earth-stabilizing-temperature-1116.
14. Michon Scott and Rebecca Lindsey, "What's the Hottest Earth's Ever Been?" Climate.gov, National Oceanic and Atmospheric Administration (hereafter NOAA), Feb. 26, 2025, https://www.climate.gov/news-features/climate-qa/whats-hottest-earths-ever-been.
15. Jun Korenaga, "Hadean Geodynamics and the Nature of Early Continental Crust," *Precambrian Research* 359 (July 2021): 106178, https://doi.org/10.1016/j.precamres.2021.106178.
16. William P. Patterson et al., "Two Millennia of North Atlantic Seasonality and Implications for Norse Colonies," *Proceedings of the National Academy of Sciences* 107, no. 12 (Mar. 2010): 5306–10, https://doi.org/10.1073/pnas.0902522107; "How Can Pollen Teach Us About Climate?" National Centers for Environmental Information (hereafter NCEI), July 27, 2016, https://www.ncei.noaa.gov/news/how-can-pollen-teach-us-about-climate.
17. Emily J. Judd et al., "A 485-Million-Year History of Earth's Surface Temperature," *Science* 385, no. 6715 (Sept. 2024): 4, https://doi.org/10.1126/science.adk3705.
18. "Great Lakes Ecoregion," NOAA, last updated Feb. 25, 2025, https://www.noaa.gov/education/resource-collections/freshwater/great-lakes-ecoregion.

19. Susan Wendell, "Toward a Feminist Theory of Disability," *Hypatia* 4, no. 2 (1989): 104–24, https://doi.org/10.1111/j.1527-2001.1989.tb00576.x.
20. James R. Fleming, "Joseph Fourier, the 'Greenhouse Effect,' and the Quest for a Universal Theory of Terrestrial Temperatures," *Endeavour* 23, no. 2 (Jan. 1999): 72–75, https://doi.org/10.1016/S0160-9327(99)01210-7.
21. Elisabeth Crawford, "Arrhenius' 1896 Model of the Greenhouse Effect in Context," *Ambio* 26, no. 1 (1997): 6–11, https://www.jstor.org/stable/4314543.
22. European Environment Agency, https://www.eea.europa.eu/data-and-maps/daviz/atmospheric-concentration-of-carbon-dioxide-5/download.table.
23. M. C. Eppes et al., "Warmer, Wetter Climates Accelerate Mechanical Weathering in Field Data, Independent of Stress-Loading," *Geophysical Research Letters* 47, no. 24 (2020): 2020GL089062, https://doi.org/10.1029/2020GL089062.
24. "TimesMachine: Tuesday October 18, 1983," *New York Times*, https://timesmachine.nytimes.com/timesmachine/1983/10/18/issue.html.
25. "Scientific Consensus," NASA Science, June 15, 2022, https://science.nasa.gov/climate-change/scientific-consensus/.
26. Jack Wall et al., "Ultrasonographic Findings in Common Thyroid and Parathyroid Disorders—Advantages of Real Time Observation by the Endocrinologist with Their Own Ultrasound Machine," *Reports* 4, no. 1 (Mar. 2021): 8, https://doi.org/10.3390/reports4010008.
27. Len Fisher, "Are Black Holes Hot or Cold?" *BBC Science Focus*, Sept. 22, 2009, https://www.sciencefocus.com/space/are-black-holes-hot-or-cold; NASA, "What Are Black Holes?" Sept. 8, 2020, https://www.nasa.gov/universe/what-are-black-holes/.
28. Wilmar M. Wiersinga, "T4+T3 Combination Therapy: An Unsolved Problem of Increasing Magnitude and Complexity," *Endocrinology and Metabolism* 36, no. 5 (Oct. 2021): 938–51, https://doi.org/10.3803/EnM.2021.501.
29. Stefan Slater, "The Discovery of Thyroid Replacement Therapy," *James Lind Library Bulletin* (2010), https://www.jameslindlibrary.org/articles/the-discovery-of-thyroid-replacement-therapy/.

HEATWAVES, HORMONES, AND SELF CONTROL

1. "What Do Volcanoes Have to Do with Climate Change?" NASA Science, Apr. 29, 2015, https://science.nasa.gov/climate-change/faq/what-do-volcanoes-have-to-do-with-climate-change/.
2. NASA, "Global Temperature," *Climate Change: Vital Signs of the Planet*, 2022, https://climate.nasa.gov/vital-signs/global-temperature?intent=121.
3. Kara J. Connelly, Julie J. Park, and Stephen H. LaFranchi, "History of the Thyroid," *Hormone Research in Paediatrics* 95, no. 6 (Nov. 2022): 546–56, https://doi.org/10.1159/000526621.
4. Hallie Maxwell, "Metamorphosis in Amphibians and the Role of Thyroid Hormone," *Cardinal Edge* 1, no. 1 (2021), https://doi.org/10.18297/tce/vol1/iss1/20.
5. Ivan Lazcano et al., "Differential Effects of 3,5-T2 and T3 on the Gill Regeneration and Metamorphosis of the Ambystoma Mexicanum (Axolotl)," *Frontiers in Endocrinology* 14 (July 2023), https://doi.org/10.3389/fendo.2023.1208182; Anne Crowner et al., "Rediscovering the Axolotl as a Model for Thyroid Hormone Dependent Development," *Frontiers in Endocrinology* 10 (Apr. 2019), https://www.frontiersin.org/journals/endocrinology/articles/10.3389/fendo.2019.00237/full.

6. Larissa V. Ponomareva et al., "Using Ambystoma Mexicanum (Mexican Axolotl) Embryos, Chemical Genetics, and Microarray Analysis to Identify Signaling Pathways Associated with Tissue Regeneration," *Comparative Biochemistry and Physiology Part C: Toxicology & Pharmacology* 178 (Dec. 2015): 128–35, https://doi.org/10.1016/j.cbpc.2015.06.004.
7. Sarah Walker, Tiago Santos-Ferreira, and Karen Echeverri, "A Reproducible Spinal Cord Crush Injury in the Regeneration-Permissive Axolotl," in *Axon Regeneration: Methods and Protocols*, ed. Ava J. Udvadia and James B. Antczak (New York: Springer US, 2023), 237–46, https://doi.org/10.1007/978-1-0716-3012-9_13.
8. Deanna H. Olson and Daniel Saenz, "Amphibians and Climate Change," Climate Change Resource Center, Dec. 12, 2023, https://www.climatehubs.usda.gov/sites/default/files/Amphibians-and-Climate%20Change_Climate-Change-Resource-Center.pdf.
9. W. A. Manasee, T. Weerathunga, and Gayani Rajapaksa, "The Impact of Elevated Temperature and CO2 on Growth, Physiological and Immune Responses of *Polypedates cruciger* (Common Hourglass Tree Frog)," *Frontiers in Zoology* 17, no. 1 (Jan. 2020): 3, https://doi.org/10.1186/s12983-019-0348-3.
10. Shogo Hori and Osamu Saitoh, "Unique High Sensitivity to Heat of Axolotl TRPV1 Revealed by the Heterologous Expression System," *Biochemical and Biophysical Research Communications* 521, no. 4 (Jan. 2020): 914–20, https://doi.org/10.1016/j.bbrc.2019.10.203.
11. Luis Zambrano, Elsa Valiente, and M. Jake Vander Zanden, "Food Web Overlap Among Native Axolotl (*Ambystoma mexicanum*) and Two Exotic Fishes: Carp (*Cyprinus carpio*) and Tilapia (*Oreochromis niloticus*) in Xochimilco, Mexico City," *Biological Invasions* 12, no. 9 (Sept. 2010): 3061–69, https://doi.org/10.1007/s10530-010-9697-8.
12. Aminetth Sánchez, "Scientists and Farmers Restore Aztec-Era Floating Farms That House Axolotls," Mongabay Environmental News, May 31, 2024, https://news.mongabay.com/2024/05/scientists-and-farmers-restore-aztec-era-floating-farms-that-house-axolotls/; Tina Deines, "Mexico City's Endangered Axolotl Has Found Fame—Is That Enough to Save It?" *National Geographic*, Apr. 25, 2025, https://www.nationalgeographic.com/animals/article/mexico-is-finally-embracing-its-quirky-salamander-the-axolotl.
13. IUCN Red List of Threatened Species, "*Ambystoma mexicanum* (Axolotl)," 2020, https://www.iucnredlist.org/species/1095/53947343.
14. Lene Andersen, "Chronic Disease in Women: Deadly and Costly," HealthCentral, Nov. 14, 2019, https://www.healthcentral.com/article/chronic-disease-in-women-deadly-and-costly; Molly Murray, "Guest Blog: A Major Health Crisis: The Alarming Rise of Autoimmune Disease," National Health Council (blog), Mar. 28, 2024, https://nationalhealthcouncil.org/blog/a-major-health-crisis-the-alarming-rise-of-autoimmune-disease/.
15. Emily Paulsen, "Recognizing, Addressing Unintended Gender Bias in Patient Care," Duke Health Referring Physicians, Jan. 14, 2020, https://physicians.dukehealth.org/articles/recognizing-addressing-unintended-gender-bias-patient-care; National Institute on Aging, "Women Hospitalized for a Heart Attack Are Less Likely to Receive Treatment and More Likely to Die Than Men," Sept. 19, 2024, https://www.nia.nih.gov/news/women-hospitalized-heart-attack-are-less-likely-receive-treatment-and-more-likely-die-men.

16. Julia C. Becker, Lea Hartwich, and S. Alexander Haslam, "Neoliberalism Can Reduce Well-Being by Promoting a Sense of Social Disconnection, Competition, and Loneliness," *British Journal of Social Psychology* 60, no. 3 (2021): 947–65, https://doi.org/10.1111/bjso.12438; Kiffer G. Card and Kirk J. Hepburn, "Is Neoliberalism Killing Us? A Cross Sectional Study of the Impact of Neoliberal Beliefs on Health and Social Wellbeing in the Midst of the COVID-19 Pandemic," *International Journal of Social Determinants of Health and Health Services* 53, no. 3 (July 2023): 363–73, https://doi.org/10.1177/00207314221134040.
17. Matthew Sparke, "Health and the Embodiment of Neoliberalism: Pathologies of Political Economy from Climate Change and Austerity to Personal Responsibility," in *Handbook of Neoliberalism*, ed. Simon Springer, Kean Birch, and Julie MacLeavy (New York: Routledge, 2016), 265–79, https://doi.org/10.4324/9781315730660-31.
18. Dan Byrne, "The 20 Most Polluting Companies in the World," Corporate Governance Institute, https://www.thecorporategovernanceinstitute.com/insights/news-analysis/the-20-most-polluting-companies-in-the-world-esg/, accessed May 29, 2025.
19. Benjamin Franta, "Early Oil Industry Disinformation on Global Warming," *Environmental Politics* 30, no. 4 (June 2021): 663–68, https://doi.org/10.1080/09644016.2020.1863703.
20. Charlotte Taylor, "Defense, Denial, and Disinformation: Uncovering the Oil Industry's Early Knowledge of Climate Change," *Common Home* (blog), Oct. 25, 2023, https://commonhome.georgetown.edu/topics/climateenergy/defense-denial-and-disinformation-uncovering-the-oil-industrys-early-knowledge-of-climate-change/.
21. Climate Files, "1988 Exxon Memo on the Greenhouse Effect," Aug. 3, 1988, https://www.climatefiles.com/exxonmobil/566/.
22. Courtney Lindwall, "IPCC Climate Change Reports: Why They Matter to Everyone on the Planet," April 14, 2023, NRDC, https://www.nrdc.org/stories/ipcc-climate-change-reports-why-they-matter-everyone-planet.
23. Kari Marie Norgaard, *Living in Denial: Climate Change, Emotions, and Everyday Life* (Cambridge, MA: MIT Press, 2011), 202.
24. Dimitrios Gounaridis and Joshua P. Newell, "The Social Anatomy of Climate Change Denial in the United States," *Scientific Reports* 14, no. 1 (Feb. 2024): 2097, https://doi.org/10.1038/s41598-023-50591-6.
25. Anthony Leiserowitz et al., *Climate Change in the American Mind: Beliefs & Attitudes*, Yale Program on Climate Change Communication, Fall 2023, https://climatecommunication.yale.edu/publications/climate-change-in-the-american-mind-beliefs-attitudes-fall-2023/.
26. Copernicus, "Copernicus: 2024 Is the First Year to Exceed 1.5°C Above Pre-Industrial Level," Jan. 10, 2025, https://climate.copernicus.eu/copernicus-2024-first-year-exceed-15degc-above-pre-industrial-level.
27. Gillian MacNaughton and A. Kayum Ahmed, "Economic Inequality and the Right to Health: On Neoliberalism, Corporatization, and Coloniality," *Health and Human Rights Journal*, May 12, 2023, https://www.hhrjournal.org/2023/12/05/economic-inequality-and-the-right-to-health-on-neoliberalism-corporatization-and-coloniality/.
28. Ronald Labonté and David Stuckler, "The Rise of Neoliberalism: How Bad Economics Imperils Health and What to Do About It," *Journal of Epidemiology and Community Health* 70, no. 3 (2016): 312–18.

29. Susan Opotow and Leah Weiss, "New Ways of Thinking About Environmentalism: Denial and the Process of Moral Exclusion in Environmental Conflict," *Journal of Social Issues* 56, no. 3 (2000): 475–90, https://doi.org/10.1111/0022-4537.00179.
30. Opotow and Weiss, "New Ways of Thinking About Environmentalism," 488.
31. Jørgen Peder Steffensen et al., "High-Resolution Greenland Ice Core Data Show Abrupt Climate Change Happens in Few Years," *Science* 321, no. 5889 (Aug. 2008): 680–84, https://doi.org/10.1126/science.1157707; Union of Concerned Scientists, "Abrupt Climate Change," July 9, 2004, https://www.ucs.org/resources/abrupt-climate-change; NCEI, "Abrupt Climate Change: A Paleo Perspective," Nov. 11, 2021, https://www.ncei.noaa.gov/products/paleoclimatology/paleo-perspectives/abrupt-climate-change.
32. Lamont-Doherty Earth Observatory, "Two Examples of Abrupt Climate Change," Earth Institute, Columbia University, https://ocp.ldeo.columbia.edu/res/div/ocp/arch/examples.shtml, accessed May 29, 2025.
33. James Lawrence Powell, "Premature Rejection in Science: The Case of the Younger Dryas Impact Hypothesis," *Science Progress* 105, no. 1 (Jan. 2022): 00368504211064272, https://doi.org/10.1177/00368504211064272; R. B. Firestone et al., "Evidence for an Extraterrestrial Impact 12,900 Years Ago That Contributed to the Megafaunal Extinctions and the Younger Dryas Cooling," *Proceedings of the National Academy of Sciences* 104, no. 41 (Oct. 2007): 16016–21, https://doi.org/10.1073/pnas.0706977104.
34. Mathew Stewart, W. Christopher Carleton, and Huw S. Groucutt, "Climate Change, Not Human Population Growth, Correlates with Late Quaternary Megafauna Declines in North America," *Nature Communications* 12, no. 1 (Feb. 2021): 965, https://doi.org/10.1038/s41467-021-21201-8.
35. David J. Meltzer, "Pleistocene Overkill and North American Mammalian Extinctions," *Annual Review of Anthropology* 44 (Oct. 2015): 33–53, https://doi.org/10.1146/annurev-anthro-102214-013854.
36. Kunio Kaiho, "Extinction Magnitude of Animals in the Near Future," *Scientific Reports* 12, no. 1 (Nov. 2022): 19593, https://doi.org/10.1038/s41598-022-23369-5.

DEATH VALLEY AND DEVILS HOLE

1. National Park Service, "Weather," Death Valley National Park, last updated Jan. 27, 2023, https://www.nps.gov/deva/learn/nature/weather-and-climate.htm.
2. National Park Service, "Endemic Plants and Animals," Death Valley National Park, last updated Feb. 8, 2021, https://www.nps.gov/deva/learn/nature/endemic-plants-and-animals.htm; US Geological Survey (USGS), "Ecology of Death Valley National Park," Geology and Ecology of National Parks, https://www.usgs.gov/geology-and-ecology-of-national-parks/ecology-death-valley-national-park-0, accessed May 29, 2025.
3. National Park Service, "Hottest Summer in Death Valley History," Death Valley National Park, Sept. 5, 2024, https://www.nps.gov/deva/learn/news/2024-hottest-dv-summer.htm.
4. Steven J. Crum, "Pretending They Didn't Exist: The Timbisha Shoshone Tribe of Death Valley, California and the Death Valley National Monument up to 1933," *Southern California Quarterly* 84, nos. 3–4 (2002): 223–40, https://doi.org/10.2307/41172134.
5. Catherine S. Fowler, "Applied Ethnobiology and Advocacy: A Case Study from the Timbisha Shoshone Tribe of Death Valley, California," *Journal of Ethnobiology* 39, no. 1 (Apr. 2019): 76–89, https://doi.org/10.2993/0278-0771-39.1.76.

6. Pauline Esteves, preface to *Draft Secretarial Report to Congress*, Apr. 1999, in Theodore Catton, *To Make a Better Nation: An Administrative History of the Timbisha Shoshone Homeland Act* (Missoula, MT: University of Montana Printing and Graphic Services, 2009), viii.
7. Paul J. White, "Troubled Waters: Timbisha Shoshone, Miners, and Dispossession at Warm Spring," *IA: The Journal of the Society for Industrial Archeology* 32, no. 1 (2006): 4–24.
8. Kim Stringfellow, "How the Timbisha Shoshone Got Their Land Back," *The Mojave Project* (blog), July 2016, https://mojaveproject.org/dispatches-item/how-the-timbisha-got-their-land-back/.
9. William Lewis Manly, *Death Valley in '49* (San Jose, CA: Pacific Tree and Vine Co., 1894), available at https://www.gutenberg.org/cache/epub/12236/pg12236-images.html.
10. Manly, *Death Valley in '49*.
11. Gregory W. Kirschen et al., "Relationship Between Body Temperature and Heart Rate in Adults and Children: A Local and National Study," *American Journal of Emergency Medicine* 38, no. 5 (May 2020): 929–33, https://doi.org/10.1016/j.ajem.2019.158355.
12. Stephanie Pappas, "How Heat Affects the Mind," American Psychological Association, June 2024, https://www.apa.org/monitor/2024/06/heat-affects-mental-health.
13. Hayon Michelle Choi et al., "Temperature, Crime, and Violence: A Systematic Review and Meta-analysis," *Environmental Health Perspectives* 132, no. 10 (Oct. 2024): 106001, https://doi.org/10.1289/EHP14300; Heather R. Stevens et al., "Associations Between Violent Crime Inside and Outside, Air Temperature, Urban Heat Island Magnitude and Urban Green Space," *International Journal of Biometeorology* 68, no. 4 (2024): 661–73, https://doi.org/10.1007/s00484-023-02613-1.
14. NOAA, "Weather Related Fatality and Injury Statistics," National Weather Service, 2024, https://www.weather.gov/hazstat.
15. US Environmental Protection Agency, "Climate Change Indicators: Heat-Related Deaths," web update June 2024, https://www.epa.gov/climate-indicators/climate-change-indicators-heat-related-deaths.
16. Noah S. Diffenbaugh and Elizabeth A. Barnes, "Data-Driven Predictions of Peak Warming Under Rapid Decarbonization," *Geophysical Research Letters* 51, no. 23 (2024):e2024GL111832, https://doi.org/10.1029/2024GL111832; Lucas R. Vargas Zeppetello, Adrian E. Raftery, and David S. Battisti, "Probabilistic Projections of Increased Heat Stress Driven by Climate Change," *Communications Earth & Environment* 3, no. 1 (Aug. 2022): 1–7, https://doi.org/10.1038/s43247-022-00524-4.
17. Rachel H. White et al., "The Unprecedented Pacific Northwest Heatwave of June 2021," *Nature Communications* 14, no. 1 (Feb. 2023): 727, https://doi.org/10.1038/s41467-023-36289-3.
18. Roberto Vita et al., "Thyroid Vascularization Is an Important Ultrasonographic Parameter in Untreated Graves' Disease Patients," *Journal of Clinical & Translational Endocrinology* 15 (Jan. 2019): 65–69, https://doi.org/10.1016/j.jcte.2019.01.001.
19. Rahul Jadhav, Mohammad Alberawi, and Khushboo Gupta, "Graves' Disease: A Rare Fate of Hashimoto's Thyroiditis," *World Journal of Nuclear Medicine* 20, no. 1 (Oct. 2020): 102–4, https://doi.org/10.4103/wjnm.WJNM_34_20.
20. Gavin Schmidt, "Climate Models Can't Explain 2023's Huge Heat Anomaly—We Could Be in Uncharted Territory," *Nature* 627, no. 8004 (Mar. 2024): 467, https://doi.org/10.1038/d41586-024-00816-z.

21. Elizabeth Kolbert, "What's Causing the Recent Spike in Global Temperatures?" *Yale e360*, Oct. 10, 2024, https://e360.yale.edu/features/gavin-schmidt-interview.
22. İsmail K. Sağlam et al., "Best Available Science Still Supports an Ancient Common Origin of Devils Hole and Devils Hole Pupfish," *Molecular Ecology* 27, no. 4 (2018): 839–42, https://doi.org/10.1111/mec.14502; LeRoy Johnson and Jean Johnson, eds., *Escape from Death Valley: As Told by William Lewis Manly and Other '49ers* (Reno: University of Nevada Press, 1987), 157.
23. IUCN Red List of Threatened Species, "*Cyprinodon diabolis*," Aug. 27, 2014, https://www.iucnredlist.org/en.
24. Kevin Wilson, Zoom interview with author, Sept. 23, 2024.
25. İsmail K. Sağlam et al., "Best Available Science Still Supports an Ancient Common Origin of Devils Hole and Devils Hole Pupfish," *Molecular Ecology* 27, no. 4 (2018): 839–42, https://doi.org/10.1111/mec.14502.
26. United States v. Cappaert; Cappaert v. United States, No. 74-1690, cert. granted sub nom.; Nos. 74-1107; 74-1304 (9th Cir. Dec. 4, 1974; June 23, 1975), https://www.elr.info/sites/default/files/litigation/5.20494.htm.
27. Olin Feuerbacher, phone interview with author, May 5, 2024.
28. "UN Report: Nature's Dangerous Decline 'Unprecedented'; Species Extinction Rates 'Accelerating,'" *Sustainable Development* (blog), May 6, 2019, https://www.un.org/sustainabledevelopment/blog/2019/05/nature-decline-unprecedented-report/.
29. John Foster, *After Sustainability: Denial, Hope, Retrieval* (New York: Routledge, 2015).
30. Sally Weintrobe, "Moral Injury, the Culture of Uncare and the Climate Bubble," *Journal of Social Work Practice* 34, no. 4 (Oct. 2020): 352, https://doi.org/10.1080/02650533.2020.1844167.

THE MEANING OF A STORM

1. J. Donald Hughes, "Dream Interpretation in Ancient Civilizations," *Dreaming* 10, no. 1 (Mar. 2000): 7–18, https://doi.org/10.1023/A:1009447606158.
2. Carole Fritz and Gilles Tosello, "From Gesture to Myth: Artists' Techniques on the Walls of Chauvet Cave," *Palethnologie* 2015, https://doi.org/10.4000/palethnologie.876.
3. UNESCO World Heritage Centre, "Göbekli Tepe," https://whc.unesco.org/en/list/1572/.
4. Joris Peters and Klaus Schmidt, "Animals in the Symbolic World of Pre-pottery Neolithic Göbekli Tepe, South-Eastern Turkey: A Preliminary Assessment," *Anthropozoologica* 39, no. 1 (2004): 179–218, https://sciencepress.mnhn.fr/en/periodiques/anthropozoologica/39/1/les-animaux-dans-le-monde-symbolique-du-ppnb-de-gobekli-tepe-turquie-du-sud-est-premiere-evaluation.
5. Daldianus Artemidorus, *The Interpretation of Dreams Digested into Five Books by That Ancient and Excellent Philosopher, Artimedorus . . .* (London: Bernard Alsop, 1644), available at Early English Books Online, Library Digital Collections, University of Michigan, https://quod.lib.umich.edu/e/eebo/A25906.0001.001/1:9.38?rgn=div2;view=fulltext.
6. Isabel Moreira, "Dreams and Divination in Early Medieval Canonical and Narrative Sources: The Question of Clerical Control," *Catholic Historical Review* 89, no. 4 (2003): 621–42.
7. Ahmed S. BaHammam, Aljohara S. Almeneessier, and Seithikurippu R. Pandi-Perumal, "Medieval Islamic Scholarship and Writings on Sleep and Dreams,"

Annals of Thoracic Medicine 13, no. 2 (2018): 72–75, https://doi.org/10.4103/atm.ATM_162_17.

8. Bronwen Neil, "Studying Dream Interpretation from Early Christianity to the Rise of Islam," *Journal of Religious History* 40, no. 1 (2016): 44–64, https://doi.org/10.1111/1467-9809.12262.
9. Tore Nielsen and Ross Levin, "Nightmares: A New Neurocognitive Model," *Sleep Medicine Reviews* 11, no. 4 (Aug. 2007): 295–310, https://doi.org/10.1016/j.smrv.2007.03.004; Victor I. Spoormaker, Michael Schredl, and Jan van den Bout, "Nightmares: From Anxiety Symptom to Sleep Disorder," *Sleep Medicine Reviews* 10, no. 1 (Feb. 2006): 19–31, https://doi.org/10.1016/j.smrv.2005.06.001.
10. Katja Valli and Antti Revonsuo, "The Threat Simulation Theory in Light of Recent Empirical Evidence: A Review," *American Journal of Psychology* 122, no. 1 (2009): 17–38.
11. "Earth's Changeable Atmosphere," *Nature Geoscience* 9, no. 6 (June 2016): 409, https://doi.org/10.1038/ngeo2735; Daniele L. Pinti and Nicholas Arndt, "Oceans, Origin Of," in *Encyclopedia of Astrobiology*, ed. Muriel Gargaud et al. (Berlin: Springer, 2020), 1–6, https://doi.org/10.1007/978-3-642-27833-4_1098-6.
12. Guillaume Paris and Pierre-Henri Blard, "Humans Have Derailed the Earth's Climate in Just 160 Years. Here's How," World Economic Forum, Mar. 27, 2019, https://www.weforum.org/stories/2019/03/how-humans-derailed-the-earths-climate-in-just-160-years/.
13. Benjamin L. Hess, Sandra Piazolo, and Jason Harvey, "Lightning Strikes as a Major Facilitator of Prebiotic Phosphorus Reduction on Early Earth," *Nature Communications* 12, no. 1 (Mar. 2021): 1535, https://doi.org/10.1038/s41467-021-21849-2.
14. Aditi Sheshadri et al., "Midlatitude Error Growth in Atmospheric GCMs: The Role of Eddy Growth Rate," *Geophysical Research Letters* 48, no. 23 (2021): e2021GL096126, https://doi.org/10.1029/2021GL096126.
15. Russell Maddicks, "Where Bolts Strike 10 Hours a Night," BBC, Feb. 28, 2022, https://www.bbc.com/travel/article/20140912-in-venezuela-natures-most-electrifying-lightning-show.
16. Rachel I. Albrecht et al., "Where Are the Lightning Hotspots on Earth?" *Bulletin of the American Meteorological Society* 97, no. 11 (Nov. 2016): 2051–68, https://doi.org/10.1175/BAMS-D-14-00193.1.
17. Agnieszka Gautier, "The Maracaibo Beacon," Earth Science Data Systems, NASA, Apr. 19, 2021, https://www.earthdata.nasa.gov/news/feature-articles/maracaibo-beacon.
18. Roberto Buizza, "Chaos and Weather Prediction," Meteorological Training Course Lecture Series, European Centre for Medium-Range Weather Forecasts, 2002, https://www.ecmwf.int/sites/default/files/elibrary/2002/16927-chaos-and-weather-prediction.pdf.
19. Edward N. Lorenz, *The Essence of Chaos* (Seattle: University of Washington Press, 1995), 82.
20. JungKyu Rhys Lim, Brooke Fisher Liu, and Michael Egnoto, "Cry Wolf Effect? Evaluating the Impact of False Alarms on Public Responses to Tornado Alerts in the Southeastern United States," *Weather, Climate, and Society* 11, no. 3 (July 2019): 549–63, https://doi.org/10.1175/WCAS-D-18-0080.1.

BROKEN BODIES ELECTRIC

1. Sappho, "Glittering-Minded, Deathless Aphrodite," http://faculty.las.illinois.edu/rrushing/241/ewExternalFiles/Sappho.pdf.

2. "Sappho," Poetry Foundation, https://www.poetryfoundation.org/poets/sappho.
3. Joanna M. Joly and José A. Tallaj, "The Heart of the (Gray) Matter Is Not Black and White," *Circulation Research* 126, no. 6 (Mar. 2020): 765–66, https://doi.org/10.1161/CIRCRESAHA.120.316698; W. C. Aird, "Discovery of the Cardiovascular System: From Galen to William Harvey," *Journal of Thrombosis and Haemostasis* 9, supplement 1 (July 2011): 118–29, https://doi.org/10.1111/j.1538-7836.2011.04312.x; Vincent Michael Figueredo, "The Ancient Heart: What the Heart Meant to Our Ancestors," *JACC* 78, no. 9 (Aug. 2021): 957–59, https://doi.org/10.1016/j.jacc.2021.06.041.
4. Wallisa Roberts et al., "Across the Centuries: Piecing Together the Anatomy of the Heart," *Translational Research in Anatomy* 17 (Nov. 2019): 100051, https://doi.org/10.1016/j.tria.2019.100051.
5. Roberts et al., "Across the Centuries."
6. Lisa Briggs and Jens Jakobsson, "Searching for Silphium: An Updated Review," *Heritage* 5, no. 2 (June 2022): 936–55, https://doi.org/10.3390/heritage5020051.
7. Gabriele Fragasso and Mauro Carlino, "The Origin of the Popular Iconic Heart Symbol: Fiction or Facts?" *Journal of Visual Communication in Medicine* 46, no. 4 (Oct. 2023): 192–96, https://doi.org/10.1080/17453054.2024.2330357.
8. Michael Miller, "Emotional Rescue: The Heart-Brain Connection," *Cerebrum: The Dana Forum on Brain Science* 2019 (May 2019): cer-05-19; Yoshihiro J. Akashi et al., "Takotsubo Cardiomyopathy," *Circulation* 118, no. 25 (Dec. 2008): 2754–62, https://doi.org/10.1161/CIRCULATIONAHA.108.767012.
9. Dany Spencer Adams, "What Is Bioelectricity?" *Bioelectricity* 1, no. 1 (Mar. 2019): 3–4, https://doi.org/10.1089/bioe.2019.0005; Luigi Catacuzzeno, Antonio Michelucci, and Fabio Franciolini, "The Long Journey from Animal Electricity to the Discovery of Ion Channels and the Modelling of the Human Brain," *Biomolecules* 14, no. 6 (June 2024): 684, https://doi.org/10.3390/biom14060684.
10. Andrew Lai, "The Experiment That Shocked The World," *Helix*, Aug. 2, 2017, https://www.helix.northwestern.edu/2017/08/02/the-experiment-that-shocked-the-world/.
11. Clay M. Armstrong and Stephen Hollingworth, "Na+ and K+ Channels: History and Structure," *Biophysical Journal* 120, no. 5 (Mar. 2021): 756–63, https://doi.org/10.1016/j.bpj.2021.01.013.
12. Robert Hays Cummins, "Struck by Lightning in the Bahamas," *Journal of Geological Education* 40, no. 3 (May 1992): 226–27, https://doi.org/10.5408/0022-1368-40.3.226.
13. NCEI, "U.S. Tornadoes," https://www.ncei.noaa.gov/access/monitoring/tornadoes/12/1?fatalities=true, accessed May 29, 2025.
14. Clark Merrefield, "How Tornadoes Can Exacerbate Racial Segregation in the US," *The Journalist's Resource* (blog), Apr. 26, 2022, https://journalistsresource.org/environment/tornado-racial-segregation/.
15. Funing Li et al., "Upstream Surface Roughness and Terrain Are Strong Drivers of Contrast in Tornado Potential Between North and South America," *Proceedings of the National Academy of Sciences* 121, no. 26 (June 2024): e2315425121, https://doi.org/10.1073/pnas.2315425121.
16. Douglas E. Miller et al., "Hybrid Prediction of Weekly Tornado Activity Out to Week 3: Utilizing Weather Regimes," *Geophysical Research Letters* 47, no. 9 (2020): e2020GL087253, https://doi.org/10.1029/2020GL087253; Ashton Robinson Cook et al., "The Impact of El Niño–Southern Oscillation (ENSO) on Winter and Early

Spring U.S. Tornado Outbreaks," *Journal of Applied Meteorology and Climatology* 56, no. 9 (Sept. 2017): 2455–78, https://doi.org/10.1175/JAMC-D-16-0249.1.

17. Bob Henson, "Climate Change and Tornadoes: Any Connection?" Yale Climate Connections, July 19, 2021, http://yaleclimateconnections.org/2021/07/climate-change-and-tornadoes-any-connection/.
18. David M. Romps et al., "Projected Increase in Lightning Strikes in the United States Due to Global Warming," *Science* 346, no. 6211 (Nov. 2014): 851–54, https://doi.org/10.1126/science.1259100.
19. Irwin Klein and Sara Danzi, "Thyroid Disease and the Heart," *Circulation* 116, no. 15 (Oct. 2007): 1725–35, https://doi.org/10.1161/CIRCULATIONAHA.106.678326.
20. Tongxing Li, Yuhan Li, and Shenglei Dongye, "Research on Chaos Theory and Chaos in Medical Practice," *Open Access Library Journal* 11, no. 4 (Apr. 2024): 1–8, https://doi.org/10.4236/oalib.1111363.
21. Community Development Block Grant Disaster Recovery, "Hurricane Sandy," NYC Recovery, https://www.nyc.gov/site/cdbgdr/hurricane-sandy/hurricane-sandy.page, accessed May 29, 2025.
22. Javad Babaie et al., "Cardiovascular Diseases in Natural Disasters: A Systematic Review," *Archives of Academic Emergency Medicine* 9, no. 1 (May 2021): e36, https://doi.org/10.22037/aaem.v9i1.1208.
23. Samantha Maldonado, "Ten Years Ago, Occupy Sandy Didn't Just Help New Yorkers, It Redefined Disaster Response," *The City*, Oct. 28, 2022, https://www.thecity.nyc/2022/10/28/ten-years-occupy-sandy-disaster-response/.
24. Josh Ehrenberg and Yael Sverdlik, in-person interview with author, Nov. 12, 2012.
25. Rachel Schragis, in-person interview with author, Nov. 18, 2012.
26. John Grzeskowiak, in-person interview with author, Nov. 30, 2012.
27. Adam Chandler, "A Tribute to the Two Jewish Hurricane Victims: Jessie Streich-Kest and Jacob Vogelman," *Tablet*, Oct. 31, 2012, https://www.tabletmag.com/sections/news/articles/a-tribute-to-the-two-jewish-hurricane-victims.

BEFRIENDING FEAR

1. M. L. West, "'Eumelos': A Corinthian Epic Cycle?" *Journal of Hellenic Studies* 122 (2002): 109–33, https://doi.org/10.2307/3246207.
2. D. Felton, "Rejecting and Embracing the Monstrous in Ancient Greece and Rome," in *The Ashgate Research Companion to Monsters and the Monstrous*, ed. Asa Simon Mittman and Peter J. Dendle (New York: Routledge, 2016), 103–32.
3. West, "'Eumelos.'"
4. Joshua C. Bregy et al., "2500-Year Paleotempestological Record of Intense Storms for the Northern Gulf of Mexico, United States," *Marine Geology* 396 (Feb. 2018): 26–42, https://doi.org/10.1016/j.margeo.2017.09.009.
5. Danielle Venton, "Exploring Ancient Storms in Marshes, Corals, and Caves," *Proceedings of the National Academy of Sciences* 113, no. 12 (Mar. 2016): 3125–26, https://doi.org/10.1073/pnas.1600798113.
6. Nikk Ogasa, "Fossilized Lightning Bolts Reveal When Ancient Storms Struck," *Science*, Dec. 16, 2020, https://www.science.org/content/article/fossilized-lightning-bolts-reveal-when-ancient-storms-struck; Jonathan M. Castro et al., "Lightning-Induced Weathering of Cascadian Volcanic Peaks," *Earth and Planetary Science Letters* 552 (Dec. 2020): 116595, https://doi.org/10.1016/j.epsl.2020.116595.

7. Guihua Wang et al., "Ocean Currents Show Global Intensification of Weak Tropical Cyclones," *Nature* 611, no. 7936 (Nov. 2022): 496–500, https://doi.org/10.1038/s41586-022-05326-4.
8. I.-I. Lin, Iam-Fei Pun, and Chun-Chi Lien, "'Category-6' Supertyphoon Haiyan in Global Warming Hiatus: Contribution from Subsurface Ocean Warming," *Geophysical Research Letters* 41, no. 23 (2014): 8547–53, https://doi.org/10.1002/2014GL061281.
9. Tosin Thompson, "Young People's Climate Anxiety Revealed in Landmark Survey," *Nature* 597, no. 7878 (Sept. 2021): 605, https://doi.org/10.1038/d41586-021-02582-8.
10. Britt Wray, *Generation Dread: Finding Purpose in an Age of Climate Crisis* (Toronto: Knopf Canada, 2022), 19–20.
11. Chelsea Sawyers et al., "The Genetic and Environmental Structure of Fear and Anxiety in Juvenile Twins," *American Journal of Medical Genetics; Part B, Neuropsychiatric Genetics* 180, no. 3 (Apr. 2019): 204–12, https://doi.org/10.1002/ajmg.b.32714.

WITCH HUNTS

1. Adam R. Sarafian et al., "Early Accretion of Water in the Inner Solar System from a Carbonaceous Chondrite–Like Source," *Science* 346, no. 6209 (Oct. 2014): 623–26, https://doi.org/10.1126/science.1256717.
2. Carnegie Science, "How Did Earth Get Its Water?" Earth & Planets Laboratory, Apr. 22, 2025, https://carnegiescience.edu/news/how-did-earth-get-its-water.
3. Pushpendra Kumar Singh et al., "Hydrology and Water Resources Management in Ancient India," *Hydrology and Earth System Sciences* 24, no. 10 (Oct. 2020): 4691–707, https://doi.org/10.5194/hess-24-4691-2020.
4. Edgar Peltenburg, "Kissonerga-Mylouthkia, Cyprus 1976–1996," Archaeology Data Service, 2009, https://archaeologydataservice.ac.uk/archives/view/mylouthkia_ba_2009/overview.cfm; Arthur Krim, "Thirst: Water and Power in the Ancient World," *AAG Review of Books* 2, no. 3 (July 2014): 80–82, https://doi.org/10.1080/2325548X.2014.919143; Simone Mantellini et al., "Development of Water Management Strategies in Southern Mesopotamia During the Fourth and Third Millennium B.C.E.," *Geoarchaeology* 39, no. 3 (2024): 268–99, https://doi.org/10.1002/gea.21992.
5. Giulio Boccaletti, *Water: A Biography* (New York: Vintage Books, 2022), xi.
6. Sophus Helle, "Tiamat (Goddess)," *Ancient Mesopotamian Gods and Goddesses*, Oracc and the UK Higher Education Academy, 2019, https://oracc.museum.upenn.edu/amgg/listofdeities/tiamat/.
7. Mathew J. Owens et al., "The Maunder Minimum and the Little Ice Age: An Update from Recent Reconstructions and Climate Simulations," *Journal of Space Weather and Space Climate* 7 (2017): A33, https://doi.org/10.1051/swsc/2017034; Beatriz Arellano-Nava et al., "Destabilisation of the Subpolar North Atlantic Prior to the Little Ice Age," *Nature Communications* 13, no. 1 (Aug. 2022): 5008, https://doi.org/10.1038/s41467-022-32653-x; Thomas J. Crowley et al., "Volcanism and the Little Ice Age," *PAGES News* 16, no. 2 (Apr. 2008): 22–23, https://doi.org/10.22498/pages.16.2.22; "The Effects of the Little Ice Age (c. 1300–1850)," *Climate in Global Cultures & Histories: Promoting Climate Literacy Across Disciplines* (blog), https://www.science.smith.edu/climatelit/the-effects-of-the-little-ice-age/, accessed May 29, 2025.
8. William Chester Jordan, "The Great Famine: 1315–1322 Revisited," in *Ecologies and Economies in Medieval and Early Modern Europe*, ed. Scott G. Bruce (Leiden. Netherlands: Brill, 2010), 44–61.

9. Guest Blogger, "Europe's 'Great Famine' Years Were Some of the Soggiest in Centuries," Lamont-Doherty Earth Observatory, Columbia Climate School, Sept. 22, 2020, https://lamont.columbia.edu/news/europes-great-famine-years-were-some-soggiest-centuries.
10. Melissa Kane, "Witches by Weather: The Impact of Climate in Early Modern Witch Trials," *Retrospect Journal* (blog), Oct. 17, 2021, https://retrospectjournal.com/2021/10/17/witches-by-weather-the-impact-of-climate-in-early-modern-witch-trials/.
11. Robert Walinski-Kiehl, "Pamphlets, Propaganda and Witch-Hunting in Germany c. 1560–c. 1630," *Reformation* 6, no. 1 (Jan. 2002): 49–74, https://doi.org/10.1179/ref_2002_6_1_004; Natalie Grace, "'Vermin and Devil-Worshippers': Exploring Witch Identities in Popular Print in Early Modern Germany and England," *Midland Historical Review*, Feb. 4, 2021, https://www.midlandshistoricalreview.com/vermin-and-devil-worshippers-exploring-witch-identities-in-popular-print-in-early-modern-germany-and-england/.
12. Philippa Carter, "Work, Gender and Witchcraft in Early Modern England," *Gender & History* 37, no. 1 (2025): 91–108, https://doi.org/10.1111/1468-0424.12717; "Witch-Hunts in Early Modern Europe (circa 1450–1750)," *Gendercide* (blog), https://www.gendercide.org/case_witchhunts.html, accessed May 29, 2025.
13. Stuart Macdonald, "Torture and the Scottish Witch-Hunt: A Re-examination," *International Review of Scottish Studies* 27 (2002), https://doi.org/10.21083/irss.v27i0.199; Rachel Pacini, "Making a Witch: The Triumph of Demonology over Popular Magic Beliefs in Early Modern Europe," Apr. 28, 2017, https://digitalcommons.usf.edu/honorstheses/191/.
14. "Witch-Hunts in Early Modern Europe."
15. Daniel W. G. Ade, "Lightning in the Folklife and Religion of the Central Andes," *Anthropos* 78, nos. 5–6 (1983): 770–88.
16. Susannah Clinker, "The Inca Capacocha Ritual: Sacred, Social and Political Ties to Landscape," *Fields/Terrains* 12 (2022): 9–13; Dagmara M. Socha et al., "Ritual Drug Use During Inca Human Sacrifices on Ampato Mountain (Peru): Results of a Toxicological Analysis," *Journal of Archaeological Science: Reports* 43 (June 2022): 103415, https://doi.org/10.1016/j.jasrep.2022.103415; Jennifer L. Faux, "Hail the Conquering Gods: Ritual Sacrifice of Children in Inca Society," *Journal of Contemporary Anthropology* 3, no. 1 (2012), https://docs.lib.purdue.edu/cgi/viewcontent.cgi?article=1017&context=jca.
17. Dagmara M. Socha, Johan Reinhard, and Ruddy Chávez Perea, "Inca Human Sacrifices on Misti Volcano (Peru)," *Latin American Antiquity* 32, no. 1 (Mar. 2021): 138–53, https://doi.org/10.1017/laq.2020.78.
18. Eve Judith Lowenstein, "Paleodermatoses: Lessons Learned from Mummies," *Journal of the American Academy of Dermatology* 50, no. 6 (June 2004): 919–36, https://doi.org/10.1016/S0190-9622(03)00914-9; Maria Constanza Ceruti, "Frozen Mummies from Andean Mountaintop Shrines: Bioarchaeology and Ethnohistory of Inca Human Sacrifice," *BioMed Research International* 2015 (2015): 439428, https://doi.org/10.1155/2015/439428; Alberto Gómez-Carballa et al., "The Complete Mitogenome of a 500-Year-Old Inca Child Mummy," *Scientific Reports* 5, no. 1 (Nov. 2015): 16462, https://doi.org/10.1038/srep16462.
19. Johan Reinhard, *The Ice Maiden: Inca Mummies, Mountain Gods, and Sacred Sites in the Andes* (Washington, DC: National Geographic Society, 2006), 27.

20. Ashley Strickland, "The Face of Juanita, the 'Ice Maiden' Mummy, Has Been Revealed," CNN, Nov. 3, 2023, https://edition.cnn.com/2023/11/03/world/ice-maiden-juanita-facial-reconstruction-scn/index.html.
21. Jennifer Nalewicki, "See How an Incan 'Ice Maiden' Comes Alive in This Step-by-Step Guide to Creating a Facial Approximation," *Live Science*, Nov. 5, 2023, https://www.livescience.com/archaeology/see-how-an-incan-ice-maiden-comes-alive-in-this-step-by-step-guide-to-creating-a-facial-approximation.
22. World Health Organization, "Drinking-Water," Sept. 13, 2023, https://www.who.int/news-room/fact-sheets/detail/drinking-water.
23. Kris Smith, "The Unequal Impacts of Flooding," Headwaters Economics, Oct. 2, 2023, https://headwaterseconomics.org/natural-hazards/unequal-impacts-of-flooding/.
24. Asako Okai, "Women Are Hit Hardest in Disasters, So Why Are Responses Too Often Gender-Blind?" United Nations Development Programme (UNDP), Mar. 24, 2022, https://www.undp.org/blog/women-are-hit-hardest-disasters-so-why-are-responses-too-often-gender-blind.
25. Laura Butterbaugh, "Why Did Hurricane Katrina Hit Women So Hard?" *Off Our Backs* 35, nos. 9–10 (2005): 17–19.
26. Butterbaugh, "Why Did Hurricane Katrina Hit Women So Hard?"
27. Marc Giudici, "Media Ethics: Coverage of Hurricane Katrina," *Media Psychology Review*, June 22, 2008, https://mprcenter.org/review/giudici-katrina-media-ethics/.
28. Alyssa Mari Thurston, Heidi Stöckl, and Meghna Ranganathan, "Natural Hazards, Disasters and Violence Against Women and Girls: A Global Mixed-Methods Systematic Review," *BMJ Global Health* 6, no. 4 (May 2021): e004377, https://doi.org/10.1136/bmjgh-2020-004377.
29. Julie A. Schumacher et al., "Intimate Partner Violence and Hurricane Katrina: Predictors and Associated Mental Health Outcomes," *Violence and Victims* 25, no. 5 (2010): 588–603.
30. Avis Jones-DeWeever, *Women in the Wake of the Storm: Examining the Post-Katrina Realities of the Women of New Orleans and the Gulf Coast*, Institute for Women's Policy Research, 2008, https://iwpr.org/wp-content/uploads/2020/11/D481.pdf.
31. Bruce Alpert, "George W. Bush Never Recovered Politically from Katrina," Nola.com, Aug. 28, 2015, https://www.nola.com/news/george-w-bush-never-recovered-politically-from-katrina/article_9b0ff883-2078-5662-8e6b-b8296249a161.html; Douglas Brinkley, "The Flood That Sank George W. Bush," *Vanity Fair*, Aug. 26, 2015, https://www.vanityfair.com/news/2015/08/hurricane-katrina-george-w-bush-new-orleans.
32. Lisa Powell, "Hurricane Ike 10 Years Later: In 2008, Powerful Storm Wiped Out Power Across the Miami Valley Region," *Dayton Daily News*, Sept. 14, 2021, https://www.daytondailynews.com/news/local/hurricane-ike-decade-ago-powerful-storm-wiped-out-power-across-the-miami-valley-region/gTnWXAm54iMGvdLITurt9M/.
33. James R. Averill, "Studies on Anger and Aggression: Implications for Theories of Emotion," *American Psychologist* 38, no. 11 (1983): 1145–60, https://doi.org/10.1037/0003-066X.38.11.1145, 1150.
34. Talia Miriam Master, "The Link Between Moral Anger and Social Activism: An Exploratory Study," PhD diss., May 2009, Rutgers, https://rucore.libraries.rutgers.edu/rutgers-lib/26136/PDF/1/play/.

35. Lynn E. Delisi, "The Katrina Disaster and Its Lessons," *World Psychiatry* 5, no. 1 (Feb. 2006): 3–4; Donald P. Moynihan, "The Response to Hurricane Katrina," International Risk Governance Council, https://citeseerx.ist.psu.edu/document?repid=rep1&type=pdf&doi=1afcfecb50d9e95433e8bc9e6df823e976a45a71, accessed May 29, 2025.
36. Jordan Heller, "An Oral History of Throwing Shoes at George W. Bush, 15 Years Later," *Intelligencer*, Dec. 14, 2023, https://nymag.com/intelligencer/article/george-bush-shoe-throw-oral-history-al-zaidi.html.
37. Matthew Weaver, "Court Cuts Iraqi Shoe Thrower's Prison Sentence," *The Guardian*, Apr. 7, 2009, https://www.theguardian.com/global/2009/apr/07/shoe-thrower-sentence-cut.
38. Martin Chulov, "Freed Iraqi Shoe Thrower Tells of Torture in Jail," *The Guardian*, Sept. 15, 2009, https://www.theguardian.com/world/2009/sep/15/iraqi-shoe-thrower-freed.
39. White House, "President Discusses Hurricane Relief in Address to the Nation," Sept. 15, 2005, https://georgewbush-whitehouse.archives.gov/news/releases/2005/09/20050915-8.html.
40. "Bush Mocked as He Returns to New Orleans," DW, Aug. 28, 2015, https://www.dw.com/en/protesters-mock-bush-on-return-to-new-orleans/a-18680580.
41. Jeffrey S. Sartin, "J. Marion Sims, the Father of Gynecology: Hero or Villain?" *Southern Medical Journal* 97, no. 5 (May 2004): 500–505, https://doi.org/10.1097/00007611-200405000-00017.
42. Nikki Rojas, "How Lucy, Betsey, and Anarcha Became Foremothers of Gynecology," *Harvard Gazette*, Mar. 30, 2023, https://news.harvard.edu/gazette/story/2023/03/how-lucy-betsey-and-anarcha-became-foremothers-of-gynecology/; Lee Boomer, "Life Story: Anarcha, Betsy, and Lucy," *Women & the American Story* (blog), https://wams.nyhistory.org/a-nation-divided/antebellum/anarcha-betsy-lucy/, accessed May 29, 2025.
43. Sara Spettel and Mark Donald White, "The Portrayal of J. Marion Sims' Controversial Surgical Legacy," *Journal of Urology* (June 2011), https://doi.org/10.1016/j.juro.2011.01.077.
44. L. L. Wall, "The Medical Ethics of Dr. J. Marion Sims: A Fresh Look at the Historical Record," *Journal of Medical Ethics* 32, no. 6 (June 2006): 346–50, https://doi.org/10.1136/jme.2005.012559.
45. Elinor Cleghorn, *Unwell Women: Misdiagnosis and Myth in a Man-Made World* (New York: Dutton, 2021).
46. Cleghorn, *Unwell Women*, 253.
47. "The Puerto Rico Pill Trials," *American Experience*, PBS, https://www.pbs.org/wgbh/americanexperience/features/pill-puerto-rico-pill-trials/, accessed May 29, 2025.
48. Natasha Furtado Dalomba, "The Birth Control Movement: The Part Played by Eugenics and Racism," Brown Pediatrics Residency, Mar. 27, 2024, https://brownpedsresidency.org/the-birth-control-movement-the-part-played-by-eugenics-and-racism/.
49. R. Sánchez-Rivera, "Coloniality and Reproductive Coercion in Puerto Rico in Light of the End of *Roe v. Wade*," Society for Cultural Anthropology, Oct. 3, 2022, https://culanth.org/fieldsights/coloniality-and-reproductive-coercion-in-puerto-rico-in-light-of-the-end-of-roe-v-wade.

WHEREVER WATER GOES

1. James Schroeder, "Ice Age Floods," National Park Service, Apr. 4, 2023, https://www.nps.gov/articles/000/ice-age-floods.htm.
2. USGS, "Study Confirms Age of Oldest Fossil Human Footprints in North America," Oct. 5, 2023, https://www.usgs.gov/news/national-news-release/study-confirms-age-oldest-fossil-human-footprints-north-america; Summer K. Praetorius et al., "Ice and Ocean Constraints on Early Human Migrations into North America Along the Pacific Coast," *Proceedings of the National Academy of Sciences* 120, no. 7 (Feb. 2023): e2208738120, https://doi.org/10.1073/pnas.2208738120.
3. USGS, "Study Confirms Age of Oldest Fossil Human Footprints in North America"; Praetorius et al., "Ice and Ocean Constraints on Early Human Migrations."
4. Mischa Haas et al., "Roman-Driven Cultural Eutrophication of Lake Murten, Switzerland," *Earth and Planetary Science Letters* 505 (Jan. 2019): 110–17, https://doi.org/10.1016/j.epsl.2018.10.027.
5. Hugo Delile et al., "Lead in Ancient Rome's City Waters," *PNAS* 11, no. 18 (Apr. 2014), https://doi.org/10.1073/pnas.1400097111.
6. François Jarrige and Thomas Le Roux, *The Contamination of the Earth: A History of Pollutions in the Industrial Age*, trans. Janice Egan and Michael Egan (Cambridge, MA: MIT Press, 2020), 20–21.
7. David Fowler et al., "A Chronology of Global Air Quality," *Philosophical Transactions of the Royal Society A: Mathematical, Physical and Engineering Sciences* 378, no. 2183 (Sept. 2020): 20190314, https://doi.org/10.1098/rsta.2019.0314; Peter N. Stearns, *The Industrial Revolution in World History*, 4th ed. (Boulder, CO: Westview Press, 2013), 173–74.
8. Andrew Hurley, "Creating Ecological Wastelands: Oil Pollution in New York City, 1870–1900," *Journal of Urban History* 20, no. 3 (May 1994), https://journals.sagepub.com/doi/10.1177/009614429402000303, 348.
9. EPA, *Population Surrounding 1,881 Superfund Sites*, Sept. 29, 2015, last updated Oct. 8, 2024, https://www.epa.gov/superfund/population-surrounding-1881-superfund-sites.
10. Collaborative for Health & Environment, "'The River Caught Fire': The Cuyahoga River Fire of 1969," https://www.healthandenvironment.org/resources/resource-library/eh-history/the-cuyahoga-river-fire-of-1969, accessed May 29, 2025.
11. Lorraine Boissoneault, "The Cuyahoga River Caught Fire at Least a Dozen Times, but No One Cared Until 1969," *Smithsonian*, June 19, 2019, https://www.smithsonianmag.com/history/cuyahoga-river-caught-fire-least-dozen-times-no-one-cared-until-1969-180972444/.
12. Keene Kelderman et al., *The Clean Water Act at 50: Promises Half Kept at the Half-Century Mark*, Environmental Integrity Project, Mar. 17, 2022, https://environmentalintegrity.org/wp-content/uploads/2022/03/CWA@50-report-EMBARGOED-3.17.22.pdf.
13. Charles W. Schmidt, "TSCA 2.0: A New Era in Chemical Risk Management," *Environmental Health Perspectives* 124, no. 10 (Oct. 2016): A182–86, https://doi.org/10.1289/ehp.124-A182.
14. National Institute of Environmental Health Sciences, "Perfluoroalkyl and Polyfluoroalkyl Substances (PFAS)," last reviewed May 6, 2025, https://www.niehs.nih.gov/health/topics/agents/pfc; Illinois Department of Public Health, "PFAS in Drinking Water," https://dph.illinois.gov/topics-services/environmental-health-protection/private-water/fact-sheets/pfas-drinking-water.html, accessed May 29, 2025.

15. USGS, "Tap Water Study Detects PFAS 'Forever Chemicals' Across the US," June 23, 2023, https://www.usgs.gov/news/national-news-release/tap-water-study-detects-pfas-forever-chemicals-across-us.
16. USGS "Tap Water Study Detects PFAS 'Forever Chemicals' Across the US."
17. J. F. Gottgens et al., "Long-Term GIS-Based Records of Habitat Changes in a Lake Erie Coastal Marsh," *Wetlands Ecology and Management* 6 (1998): 5–17, https://doi.org/10.1023/A:1008439519760.
18. Coastal Resilience, "Western Lake Erie," Nature Conservancy, 2021, https://coastalresilience.org/project/western-lake-erie/.
19. Department of Environment, Great Lakes, and Energy, "Great Lakes Water Levels," https://www.michigan.gov/egle/about/organization/water-resources/submerged-lands/great-lakes-water-levels, accessed May 29, 2025.
20. EPA, *Climate Change Indicators: Great Lakes Water Levels and Temperatures*, July 1, 2016, https://www.epa.gov/climate-indicators/great-lakes; Great Lakes Research Center, *Future Rise of the Great Lakes Water Levels Under Climate Change*, Michigan Technological University, 2022, https://www.mtu.edu/greatlakes/research-highlights/climate-change-great-lakes/.
21. Claire Fuschi et al., "Microplastics in the Great Lakes: Environmental, Health, and Socioeconomic Implications and Future Directions," *ACS Sustainable Chemistry & Engineering* 10, no. 43 (Oct. 2022): 14074–91, https://doi.org/10.1021/acssuschemeng.2c02896.
22. J. Bernier et al., "Immunotoxicity of Heavy Metals in Relation to Great Lakes," *Environmental Health Perspectives* 103, Supp. 9 (Dec. 1995): 23–34, https://pmc.ncbi.nlm.nih.gov/articles/PMC1518818/; Dragana Javorac et al., "Exploring the Endocrine Disrupting Potential of Lead Through Benchmark Modelling—Study in Humans," *Environmental Pollution* 316 (Jan. 2023): 120428, https://doi.org/10.1016/j.envpol.2022.120428; K. P. Mishra, "Lead Exposure and Its Impact on Immune System: A Review," *Toxicology in Vitro* 23, no. 6 (Sept. 2009): 969–72, https://doi.org/10.1016/j.tiv.2009.06.014.
23. Alexandre Vallée et al., "Pollution and Endometriosis: A Deep Dive into the Environmental Impacts on Women's Health," *BJOG: An International Journal of Obstetrics & Gynaecology* 131, no. 4 (2024): 401–14, https://doi.org/10.1111/1471-0528.17687; Vicki S. Blazer et al., "Indicators of Exposure to Estrogenic Compounds at Great Lakes Areas of Concern: Species and Site Comparisons," *Environmental Monitoring and Assessment* 190, no. 10 (2018): 577, https://doi.org/10.1007/s10661-018-6943-5.
24. Francesca Coperchini et al., "Thyroid Disrupting Effects of Old and New Generation PFAS," *Frontiers in Endocrinology* 11 (Jan. 2021): 612320, https://doi.org/10.3389/fendo.2020.612320.
25. Ariel Washington and Jill Randall, "'We're Not Taken Seriously': Describing the Experiences of Perceived Discrimination in Medical Settings for Black Women," *Journal of Racial and Ethnic Health Disparities* 10, no. 2 (Apr. 2023): 883–91, https://doi.org/10.1007/s40615-022-01276-9; Axenya Kachen and Jennifer R. Pharr, "Health Care Access and Utilization by Transgender Populations: A United States Transgender Survey Study," *Transgender Health* 5, no. 3 (Sept. 2020), https://doi.org/10.1089/trgh.2020.0017; Monika Damle, Heather Wurtz, and Goleen Samari, "Racism and Health Care: Experiences of Latinx Immigrant Women in NYC During COVID-19," *SSM: Qualitative Research in Health* 2 (Dec. 2022): 100094, https://doi.org/10.1016/j.ssmqr

.2022.100094; Kiara Tanta-Quidgeon, *Understanding How the U.S. Healthcare System Can Better Serve Indigenous People Through the Lived-Experiences of Five Indigenous Women*, United States of Care, 2023, https://unitedstatesofcare.org/wp-content/uploads/2023/06/Understanding-how-the-U.S.-Healthcare-System-can-Better-Serve-Indigenous-People-Through-the-Lived-Experiences-of-Five-Indigenous-Women-1.pdf.

26. Joanna Ebenstein, *The Anatomical Venus: Wax, God, Death & the Ecstatic* (New York, Distributed Art Publishers, 2016).
27. Thomas A. H. McCulloch, "Theories of Hysteria," *Canadian Psychiatric Association Journal* 14, no. 6 (Dec. 1969): 635–37, https://doi.org/10.1177/070674376901400614.
28. Moreno Paulon, "Hysteria: Rise and Fall of a Baffling Disease. A Review on History of Ideas in Medicine," *Journal of Psychopathology* (Sept. 2022): 152–61, https://doi.org/10.36148/2284-0249-461.
29. Helen King, "Once Upon a Text: Hysteria from Hippocrates," in *Hysteria Beyond Freud*, ed. Sander L. Gilman, Helen King, and Roy Porter (Berkeley: University of California Press, 1993), 18–19.
30. King, "Once Upon a Text," 7.
31. King, "Once Upon a Text," 4.
32. Helen King, *Hippocrates' Woman: Reading the Female Body in Ancient Greece* (London: Routledge, 1998), 14–15.
33. King, *Hippocrates' Woman*, 13.
34. King, *Hippocrates' Woman*, 17.
35. Heather A. Cahill, "Male Appropriation and Medicalization of Childbirth: An Historical Analysis," *Journal of Advanced Nursing* 33, no. 3 (Sept. 2000): 334–42; Barbara Ehrenreich and Deirdre English, *Witches, Midwives, and Nurses: A History of Women Healers* (New York: Feminist Press, 1973).
36. Anuli Njoku et al., "Listen to the Whispers Before They Become Screams: Addressing Black Maternal Morbidity and Mortality in the United States," *Healthcare* 11, no. 3 (Jan. 2023): 438, https://doi.org/10.3390/healthcare11030438.
37. Jolynn Dellinger and Stephanie Pell, "Bodies of Evidence: The Criminalization of Abortion and Surveillance of Women in a Post-*Dobbs* World," *Duke Journal of Constitutional Law & Public Policy* 19, no. 1 (Apr. 2024): 1–108, doi:10.2105/AJPH.2019.305243.
38. Carter Sherman, "Ohio Woman Won't Be Indicted for Abuse of Corpse After Miscarriage, Grand Jury Decides," *The Guardian*, Jan. 11, 2024, https://www.theguardian.com/world/2024/jan/11/ohio-woman-miscarriage-not-indicted-abuse-corpse-brittany-watts; Karine Coen-Sanchez et al., "Repercussions of Overturning *Roe v. Wade* for Women Across Systems and Beyond Borders," *Reproductive Health* 19, no. 184 (2022), https://link.springer.com/content/pdf/10.1186/s12978-022-01490-y.pdf.
39. Sherman, "Ohio Woman Won't Be Indicted for Abuse of Corpse After Miscarriage."
40. Alejandra O'Connell-Domenech, "GOP Lawmakers in 10 States Introduce Bills to Treat Abortion as Homicide," *The Hill* (blog), Mar. 28, 2025, https://thehill.com/policy/healthcare/5217297-republican-state-lawmakers-abortion-homicide-bills/.
41. Pregnancy Justice, *The Rise of Pregnancy Criminalization: A Pregnancy Justice Report*, https://www.pregnancyjusticeus.org/rise-of-pregnancy-criminalization-report/, accessed May 29, 2025.
42. Michele Goodwin, *Policing the Womb: Invisible Women and the Criminalization of Motherhood* (Cambridge: Cambridge University Press, 2020), 11.

43. Libby Stanford and Ileana Najarro, "State Laws Restricting Curriculum, Pronoun Use Cause Confusion and Chaos in Schools," *Education Week*, Sept. 21, 2023, https://www.edweek.org/policy-politics/state-laws-restricting-curriculum-pronoun-use-cause-confusion-and-chaos-in-schools/2023/09.
44. American Society of Civil Engineers, "Dams," *2025 Report Card for America's Infrastructure*, Jan. 17, 2017, https://infrastructurereportcard.org/cat-item/dams-infrastructure/.
45. Heather Randell and Andrew Curley, "Dams and Tribal Land Loss in the United States," *Environmental Research Letters* 18, no. 9 (Aug. 2023): 094001, https://doi.org/10.1088/1748-9326/acd268.

FLOOD MEMORY

1. NOAA Research, "Could Drying the Stratosphere Help Cool the Planet?" Climate.gov, Mar. 2, 2024, https://www.climate.gov/news-features/feed/could-drying-stratosphere-help-cool-planet.
2. Climate Signals, "Atmospheric Moisture Increase," https://www.climatesignals.org/climate-signals/atmospheric-moisture-increase, accessed May 29, 2025.
3. Giulio Boccaletti, *Water: A Biography* (New York: Vintage Books, 2022), 5.
4. Kevin Trenberth, "How Rising Water Vapour Is Amplifying Warming and Making Extreme Weather Worse," PreventionWeb, Sept. 13, 2023, https://www.preventionweb.net/news/how-rising-water-vapour-atmosphere-amplifying-warming-and-making-extreme-weather-worse.
5. Chi-Cherng Hong et al., "Causes of 2022 Pakistan Flooding and Its Linkage with China and Europe Heatwaves," *NPJ Climate and Atmospheric Science* 6, no. 1 (Oct. 2023): 1–10, https://doi.org/10.1038/s41612-023-00492-2; British Red Cross, "Pakistan Floods and Climate Change," last updated Aug. 30, 2023, https://www.redcross.org.uk/stories/disasters-and-emergencies/world/climate-change-and-pakistan-flooding-affecting-millions.
6. National Environmental Satellite, Data, and Information Service, "Heavy Rains Cause Flooding in New York City," Apr. 28, 2025, https://www.nesdis.noaa.gov/news/heavy-rains-cause-flooding-new-york-city; Victoria Bekiempis, "New York City Reels After Flash Flooding Chaos and Powerful Downpours," *The Guardian*, Sept. 30, 2023, https://www.theguardian.com/us-news/2023/sep/30/new-york-city-flash-flooding-aftermath.
7. Teresa Medrano and Joseph Wilson, "What to Know About the Unprecedented Floods That Killed More Than 200 in Spain," AP News, Oct. 31, 2024, https://apnews.com/article/flash-floods-spain-valencia-climate-change-what-to-know-f942142b82de24f5b4a18867bc32ae00.
8. Beverly G. Reed and Bruce R. Carr, "The Normal Menstrual Cycle and the Control of Ovulation," in *Endotext*, ed. Kenneth R. Feingold et al. (South Dartmouth, MA: MDText.com, 2000–), http://www.ncbi.nlm.nih.gov/books/NBK279054/.
9. Sarah Romans et al., "Mood and the Menstrual Cycle: A Review of Prospective Data Studies," *Gender Medicine* 9, no. 5 (Oct. 2012): 361–84, https://doi.org/10.1016/j.genm.2012.07.003.
10. G. Einstein, J. Downar, and S. H. Kennedy, "Gender/Sex Differences in Emotions," *Medicographia* 35, no. 3 (2013): 271–80.

11. Charlotte Wessel Skovlund et al., "Association of Hormonal Contraception with Depression," *JAMA Psychiatry* 73, no. 11 (Nov. 2016): 1154–62, https://doi.org/10.1001/jamapsychiatry.2016.2387.
12. Beatrice Adler-Bolton and Artie Vierkant, *Health Communism* (New York: Verso, 2022), 6–7.
13. "The Floods—10 Years On," Gloucestershire Live, July 20, 2007, http://thefloods gloucestershirelive.co.uk.
14. Matthew Weaver, "Power Station Saved as Floodwaters Recede," *The Guardian*, July 24, 2007, https://www.theguardian.com/uk/2007/jul/24/weather.world.
15. Lindsey McEwen, video interview with author, Nov. 17, 2023.
16. Ray Tobler et al., "Aboriginal Mitogenomes Reveal 50,000 Years of Regionalism in Australia," *Nature* 544, no. 7649 (Apr. 2017): 180–84, https://doi.org/10.1038/nature21416; Patrick D. Nunn, "The Oldest True Stories in the World," *Sapiens*, Oct. 18, 2018, https://www.sapiens.org/language/oral-tradition/.
17. Patrick D. Nunn and Nicholas J. Reid, "Aboriginal Memories of Inundation of the Australian Coast Dating from More Than 7000 Years Ago," *Australian Geographer* 47, no. 1 (Jan. 2016): 11–47, https://doi.org/10.1080/00049182.2015.1077539.
18. Melissa Denchak, "Flooding Facts, Causes, and Prevention," NRDC, Nov. 3, 2023, https://www.nrdc.org/stories/flooding-and-climate-change-everything-you-need-know.
19. Andrea Tone and Mary Koziol, "(F)Ailing Women in Psychiatry: Lessons from a Painful Past," *CMAJ: Canadian Medical Association Journal* 190, no. 20 (May 2018): E624–25, https://doi.org/10.1503/cmaj.171277.
20. Vanessa Zoltan, phone interview with author, Dec. 20, 2023.
21. Hilary Mantel, *Giving Up the Ghost: A Memoir* (New York, Picador: 2004), 206.
22. Mantel, *Giving Up the Ghost*, 317–18.
23. Wacław M. Adamczyk et al., "Memory of Pain in Adults: A Protocol for Systematic Review and Meta-analysis," *Systematic Reviews* 8 (Aug. 2019): 201, https://doi.org/10.1186/s13643-019-1115-4.
24. Optimal Health Lab, "The Degree of Tissue Damage Does Not Align Directly with the Level of Pain I Am Experiencing?" Dec. 5, 2023, https://www.optimalhealthlab.com.au/blogs/optimal-tips/the-degree-of-tissue-damage-does-not-align-directly-with-the-level-of-pain-i-am-experiencing.
25. Ian S. Fraser, "Mysteries of Endometriosis Pain: Chien-Tien Hsu Memorial Lecture 2009," *Journal of Obstetrics and Gynaecology Research* 36, no. 1 (Feb. 2010): 1–10, https://doi.org/10.1111/j.1447-0756.2010.01181.x.
26. Ivo Brosens and Giuseppe Benagiano, "Endometriosis, a Modern Syndrome," *Indian Journal of Medical Research* 133, no. 6 (June 2011): 581–93.
27. Annika J. Penzer and Scott J. Schweikart, "Using Policy and Law to Help Reduce Endometriosis Diagnostic Delay," *AMA Journal of Ethics* 27, no. 2 (Feb. 2025): 104–9, https://doi.org/10.1001/amajethics.2025.104; Annalise Weckesser, Danielle Perro, and Veronique Griffith, "Endometriosis: Black Women Continue to Receive Poorer Care for the Condition," *The Conversation*, Mar. 30, 2023, http://theconversation.com/endometriosis-black-women-continue-to-receive-poorer-care-for-the-condition-200663; Cheryl Eder and Rizwana Roomaney, "Transgender and Non-binary People's Experience of Endometriosis," *Journal of Health Psychology* (Aug. 2024), https://journals.sagepub.com/doi/full/10.1177/13591053241266249.

28. Perri Meldon, "Disability History: The Disability Rights Movement," National Park Service, last updated Feb. 14, 2025, https://www.nps.gov/articles/disabilityhistoryrights movement.htm.
29. "Lake Erie Bill of Rights," University of Toledo, 2019, https://www.utoledo.edu/law /academics/ligl/pdf/2019/Lake-Erie-Bill-of-Rights-GLWC-2019.pdf.
30. Apurva Lad et al., "As We Drink and Breathe: Adverse Health Effects of Microcystins and Other Harmful Algal Bloom Toxins in the Liver, Gut, Lungs and Beyond," *Life* 12, no. 3 (Mar. 2022): 418, https://doi.org/10.3390/life12030418.
31. Tish O'Dell, "The Killing of Lake Erie and Discovering Right Relationship Ten Years Later," Community Environmental Legal Defense Fund (CELDF), Aug. 6, 2024, https://celdf.org/2024/08/the-killing-of-lake-erie-and-discovering-right-relationship -ten-years-later/.
32. CELDF, "Ohio Legislature Moves to Ban Rights of Nature Enforcement," July 18, 2019, https://celdf.org/2019/07/rights-of-nature-ban/.
33. KFF, "HealthCare.Gov Insurers Denied Nearly 1 in 5 In-Network Claims in 2023, but Information About Reasons Is Limited in Public Data," news release, Jan. 27, 2025, https://www.kff.org/affordable-care-act/press-release/healthcare-gov-insurers-denied -nearly-1-in-5-in-network-claims-in-2023-but-information-about-reasons-is-limited-in -public-data/; Maya Brownstein, "Private Equity's Appetite for Hospitals May Put Patients at Risk," Harvard T. H. Chan School of Public Health, Dec. 16, 2024, https:// hsph.harvard.edu/news/private-equitys-appetite-for-hospitals-may-put-patients-at-risk/.
34. Cesar O. Estien et al., "Historical Redlining Is Associated with Disparities in Environmental Quality Across California," *Environmental Science & Technology Letters* 11, no. 2 (Jan. 2024): 54–59, https://doi.org/10.1021/acs.estlett.3c00870; Rachel Yavinsky, "Women More Vulnerable Than Men to Climate Change," Population Reference Bureau, Dec. 26, 2012, https://www.prb.org/resources/women-more-vulnerable-than -men-to-climate-change/.
35. Audre Lorde, "The Uses of Anger: Women Responding to Racism" (speech, National Women's Studies Association Conference, Storrs, CT, June 1981), published on BlackPast.org, Aug. 12, 2012, https://www.blackpast.org/african-american-history/1981 -audre-lorde-uses-anger-women-responding-racism/.
36. Barbara Deming, "On Anger," 1971, reprinted in *Peace News*, Apr. 3, 2014, https:// peacenews.info/node/7610/anger.

THE BREAKDOWN

1. Christopher H. Marvin et al., "Surficial Sediment Contamination in Lakes Erie and Ontario: A Comparative Analysis," *Journal of Great Lakes Research* 28, no. 3 (Jan. 2002): 437–50, https://doi.org/10.1016/S0380-1330(02)70596-0.
2. Dana Bogart, "'My Great Terror, the Black Swamp': Northwest Ohio's Environmental Borderland," master's thesis, Miami University, 2015, https://etd.ohiolink.edu/acprod /odb_etd/ws/send_file/send?accession=miami1429966484&disposition=inline, 11.
3. Dale Mize, "Draining the Swamp: The Destruction of an Essential Landscape," *Ohio History* 131, no. 1 (Spring 2024): 34, https://oaks.kent.edu/ohj/ohio-history-spring -2024.
4. Christopher H. Marvin et al., "Surficial Sediment Contamination in Lakes Erie and Ontario," *Journal of Great Lakes Research* 28, no. 3 (2002): 437–50, https://doi.org /10.1016/S0380-1330(02)70596-0.

5. Brian Henn et al., "Hydroclimatic Conditions Preceding the March 2014 Oso Landslide," *Journal of Hydrometeorology* 16, no. 3 (June 2015): 1243–49, https://doi.org/10.1175/JHM-D-15-0008.1.
6. Kate Allstadt, "Seismic Signals Generated by the March 22nd Oso Landslide," Pacific Northwest Seismic Network, Mar. 26, 2014, https://pnsn.org/blog/2014/03/26/seismic-signals-generated-by-the-march-22nd-oso-landslide; Dave Petley, "The Oso (Steelhead) Landslide: Mechanisms of Movement and the Challenges of Recovering the Victims," *The Landslide Blog* (blog), Mar. 28, 2014, https://blogs.agu.org/landslide blog/2014/03/28/oso-mechanisms-1/.
7. USGS, "Oso Landslide Research Paves Way for Future Hazard Evaluations," news release, Jan. 12, 2015, https://www.usgs.gov/news/national-news-release/oso-landslide-research-paves-way-future-hazard-evaluations.
8. Greg Copeland, "'None of This Is Making Sense': First Responders Recall First Impressions of Oso Landslide," King 5, Mar. 20, 2024, https://www.king5.com/article/news/local/oso-landslide/oso-landslide-rescues/281-f1ec7926-f570-400a-b44a-26d53f5b3756.
9. King 5 Staff, "Timeline: Oso, the Deadliest Landslide in US History," Mar. 18, 2024, King 5, https://www.king5.com/article/news/local/oso-landslide/timeline-oso-deadliest-landslide-us-history/281-fa64117e-5f80-4510-816c-740fbc7b08ec.
10. USGS, "1980 Cataclysmic Eruption," Nov. 7, 2023, https://www.usgs.gov/volcanoes/mount-st.-helens/science/1980-cataclysmic-eruption.
11. Richard M. Iverson, "Landslide Disparities, Flume Discoveries, and Oso Despair," *Perspectives of Earth and Space Scientists* 1, no. 1 (2020): https://doi.org/10.1029/2019CN000117.
12. R. M. Iverson et al., "Landslide Mobility and Hazards: Implications of the 2014 Oso Disaster," *Earth and Planetary Science Letters* 412 (Feb. 2015): 197–208, https://doi.org/10.1016/j.epsl.2014.12.020.
13. AP News, "Memorial at Site of Deadliest Landslide in US History Opens on 10th Anniversary," Mar. 22, 2024, https://apnews.com/article/oso-landslide-memorial-washington-2e9453ecc4658d1d8fb39ef50085e080; Randall W. Jibson, "The Mameyes, Puerto Rico, Landslide Disaster of October 7, 1985," in *Landslides/Landslide Mitigation*, vol. 9, ed. James E. Slosson, Arthur G. Keene, and Jeffrey A. Johnson (McLean, VA: Geological Society of America, 1992).
14. Courtney Wells, Zoom interview with author, Aug. 15, 2024.
15. Rebecca Gillett and Courtney Wells, "A Thousand Tiny Deaths: Attending to Grief in Patients with Rheumatic Diseases," presentation, ACR Convergence 2023, Nov. 14, 2023, https://acr23.eventscribe.net/fsPopup.asp?Mode=sessioninfo&PresentationID=1229199.
16. Quoted in Elizabeth Rush, *The Quickening: Creation and Community at the Ends of the Earth* (Minneapolis: Milkweed Editions, 2023), 215.
17. Mary Annaïse Heglar, "Climate Grief Hurts Because It's Supposed To," *The Nation*, Nov. 7, 2021, https://www.thenation.com/article/environment/climate-grief-hope/.
18. Lawrence R. Walker and Aaron B. Shiels, "Landslide Ecology," Animal and Plant Inspection Service, USDA, 2013, p. 6, https://digitalcommons.unl.edu/icwdm_usdanwrc/1644.
19. James M. Tiedje et al., "Opening the Black Box of Soil Microbial Diversity," *Applied Soil Ecology* 13, no. 2 (Oct. 1999): 109–22, https://doi.org/10.1016/S0929-1393(99)00026-8;

Roeland Cortois and Gerlinde Barbara De Deyn, "The Curse of the Black Box," *Plant and Soil* 350, no. 1 (Jan. 2012): 27–33, https://doi.org/10.1007/s11104-011-0963-z.

20. National Climate Assessment, "Introduction: Heavy Downpours Increasing," GlobalChange.gov, 2014, https://nca2014.globalchange.gov/report/our-changing-climate/heavy-downpours-increasing.
21. Nicholas Le Pan, "This Is What the Human Impact on the Earth's Surface Looks Like," World Economic Forum, Dec. 4, 2020, https://www.weforum.org/stories/2020/12/visualizing-the-human-impact-on-the-earth-s-surface/.
22. Melanie J. Froude and David N. Petley, "Global Fatal Landslide Occurrence from 2004 to 2016," *Natural Hazards and Earth System Sciences* 18, no. 8 (Aug. 2018): 2161–81, https://doi.org/10.5194/nhess-18-2161-2018.
23. Dave Petley, Zoom interview with the author, July 3, 2024.
24. Fausto Guzzetti et al., "Geographical Landslide Early Warning Systems," *Earth-Science Reviews* 200 (Jan. 2020): 102973, https://doi.org/10.1016/j.earscirev.2019.102973; Hong Kong Slope Safety, "Landslip Warning System," Protection Against Landslide Emergency, last revision dated Feb. 21, 2025, https://hkss.cedd.gov.hk/hkss/en/protection-against-landslide-emergency/landslip-warning-system/index.html.
25. Octavia Butler, *Parable of the Sower* (New York: Four Walls Eight Windows, 1993).
26. Matthew C. Larsen et al., "Natural Hazards on Alluvial Fans: The Venezuela Debris Flow and Flash Flood Disaster," Fact Sheet, USGS, 2002, https://doi.org/10.3133/fs10301.
27. Emily B. Hollister, Chunxu Gao, and James Versalovic, "Compositional and Functional Features of the Gastrointestinal Microbiome and Their Effects on Human Health," *Gastroenterology* 146, no. 6 (May 2014): 1449–58, https://doi.org/10.1053/j.gastro.2014.01.052.
28. Kathryn Harkup, "How Chilean Arsenic Eaters Vindicated a Classic Work of Crime Fiction," *The Guardian*, May 9, 2017, https://www.theguardian.com/science/blog/2017/may/09/chilean-arsenic-eaters-classic-work-of-crime-fiction-strong-poison-dorothy-l-sayers.
29. Ayman Abdo, Jonathan Meddings, and Mark Swain, "Liver Abnormalities in Celiac Disease," *Clinical Gastroenterology and Hepatology* 2, no. 2 (Feb. 2004): 107–12, https://doi.org/10.1016/S1542-3565(03)00313-6.

MICROBIAL MAYHEM

1. K. Hickman-Lewis et al., "Advanced Two- and Three-Dimensional Insights into Earth's Oldest Stromatolites (ca. 3.5 Ga): Prospects for the Search for Life on Mars," *Geology* 51, no. 1 (Nov. 2022): 33–38, https://doi.org/10.1130/G50390.1; Alexandra Witze, "Claims of Earth's Oldest Fossils Tantalize Researchers," *Nature*, Aug. 31, 2016, https://doi.org/10.1038/nature.2016.20506.
2. Richard Allen White, Pieter T. Visscher, and Brendan P. Burns, "Between a Rock and a Soft Place: The Role of Viruses in Lithification of Modern Microbial Mats," *Trends in Microbiology* 29, no. 3 (Mar. 2021): 204–13, https://doi.org/10.1016/j.tim.2020.06.004.
3. Cherie Winner, "What Doomed the Stromatolites?" *Oceanus* (blog), Nov. 15, 2013, https://www.whoi.edu/oceanus/feature/what-doomed-the-stromatolites/.
4. Dario Harazim et al., "Bioturbating Animals Control the Mobility of Redox-Sensitive Trace Elements in Organic-Rich Mudstone," *Geology* 43, no. 11 (Nov. 2015): 1007–10, https://doi.org/10.1130/G37025.1.

5. Laura G. Shields, "Did Air Pollution Cause a Deadly Chinese Landslide?" *Science*, Dec. 12, 2017, https://www.science.org/content/article/did-air-pollution-cause-deadly-chinese-landslide.
6. Ming Zhang and Mauri J. McSaveney, "Is Air Pollution Causing Landslides in China?" *Earth and Planetary Science Letters* 481 (Jan. 2018): 284–89, https://doi.org/10.1016/j.epsl.2017.10.045.
7. Zhang and McSaveney, "Is Air Pollution Causing Landslides in China?"
8. Marco Guida and Marco Borra, "Microbial Diversity of Landslide Soils Assessed by RFLP and SSCP Fingerprints," *Journal of Applied Genetics* 55 (2014): 403–15, https://doi.org/10.1007/S13353-014-0208-Y.
9. Mark A. Anthony, S. Franz Bender, and Marcel G. A. van der Heijden, "Enumerating Soil Biodiversity," *Proceedings of the National Academy of Sciences* 120, no. 33 (Aug. 2023): https://doi.org/10.1073/pnas.2304663120.
10. Max Reinold et al., "The Vulnerability of Microbial Ecosystems in a Changing Climate: Potential Impact in Shark Bay," *Life* 9, no. 3 (Sept. 2019): 71, https://doi.org/10.3390/life9030071.
11. Shanan E. Peters, Jon M. Husson, and Julia Wilcots, "The Rise and Fall of Stromatolites in Shallow Marine Environments," *Geology* 45, no. 6 (June 2017): 487–90, https://doi.org/10.1130/g38931.1.
12. Genming Luo, Deng Liu, and Hao Yang, "Microbes in Mass Extinction: An Accomplice or a Savior?" *National Science Review* 11, no. 1 (Jan. 2024): nwad291, https://doi.org/10.1093/nsr/nwad291.
13. Luo, Liu, and Yang, "Microbes in Mass Extinction."
14. Daniel F. Q. Smith and Arturo Casadevall, "Disaster Microbiology—a New Field of Study," *mBio* 13, no. 4 (Aug. 2022): e01680-22, https://doi.org/10.1128/mbio.01680-22.
15. Roberta Magnano San Lio et al., "How Antimicrobial Resistance Is Linked to Climate Change: An Overview of Two Intertwined Global Challenges," *International Journal of Environmental Research and Public Health* 20, no. 3 (Jan. 2023): 1681, https://doi.org/10.3390/ijerph20031681.
16. Elizabeth J. Archer et al., "Climate Warming and Increasing Vibrio Vulnificus Infections in North America," *Scientific Reports* 13, no. 1 (Mar. 2023): 3893, https://doi.org/10.1038/s41598-023-28247-2; Lauren Schneider, "Climate Change Is Driving Dangerous Bacteria Farther North," *Eos*, Dec. 13, 2024, https://eos.org/articles/climate-change-is-driving-dangerous-bacteria-farther-north.
17. William F. Martin and Marek Mentel, "Origin of Mitochondria," *Nature Education* 3, no. 9 (2010), https://www.nature.com/scitable/topicpage/the-origin-of-mitochondria-14232356/.
18. István Zachar and Eörs Szathmáry, "Breath-Giving Cooperation: Critical Review of Origin of Mitochondria Hypotheses," *Biology Direct* 12, no. 1 (Aug. 2017): 19, https://doi.org/10.1186/s13062-017-0190-5.
19. NIH Human Microbiome Project, "About the Human Microbiome," 2025, https://www.hmpdacc.org/overview/.
20. Danping Zheng, Timur Liwinski, and Eran Elinav, "Interaction Between Microbiota and Immunity in Health and Disease," *Cell Research* 30, no. 6 (June 2020): 492–506, https://doi.org/10.1038/s41422-020-0332-7.

21. Juliette C. Madan et al., "Effects of Cesarean Delivery and Formula Supplementation on the Intestinal Microbiome of Six-Week-Old Infants," *JAMA Pediatrics* 170, no. 3 (Mar. 2016): 212–19, https://doi.org/10.1001/jamapediatrics.2015.3732.
22. Alexis B. Dunn et al., "The Maternal Infant Microbiome: Considerations for Labor and Birth," *MCN: The American Journal of Maternal Child Nursing* 42, no. 6 (2017): 318–25, https://doi.org/10.1097/NMC.0000000000000373; Maria Isabel de Moraes-Pinto, Fabíola Suano-Souza, and Carolina S. Aranda, "Immune System: Development and Acquisition of Immunological Competence," *Jornal de Pediatria* 97, Suppl. 1 (Nov. 2020): S59–66, https://doi.org/10.1016/j.jped.2020.10.006.
23. Stanford Medicine Children's Health, "Optimal C-Section Rate May Be as High as 19 Percent," news release, Dec. 1, 2015, https://www.stanfordchildrens.org/en/about/news/releases/2015/optimal-c-section-rate-may-be-as-high-as-19-percent-to-save-lives-of-mothers-and-infants.html.
24. Jesse Anttila-Hughes et al., "Mortality from Nestlé's Marketing of Infant Formula in Low and Middle-Income Countries," National Bureau of Economic Research, Mar. 2018, https://www.nber.org/papers/w24452.
25. Johns Hopkins Medicine, "Breast Milk Is Best," Feb. 29, 2024, https://www.hopkinsmedicine.org/health/conditions-and-diseases/breastfeeding-your-baby/breast-milk-is-the-best-milk.
26. World Health Organization, "Breastfeeding," https://www.who.int/health-topics/breastfeeding, accessed May 29, 2025.
27. Giovanni Federico, "The Growth of World Agricultural Production, 1800–1938," *Research in Economic History* 22 (2004): 125–81, https://doi.org/10.1016/S0363-3268(04)22003-1.
28. Angus C. Chu, Pietro F. Peretto, and Xilin Wang, "Agricultural Revolution and Industrialization," *Journal of Development Economics* 158 (Sept. 2022): 102887, https://doi.org/10.1016/j.jdeveco.2022.102887.
29. J. L. van Zanden, "The First Green Revolution: The Growth of Production and Productivity in European Agriculture, 1870–1914," *Economic History Review* 44, no. 2 (1991): 215–39, https://doi.org/10.2307/2598294.
30. Federico, "The Growth of World Agricultural Production."
31. Danny Llewellyn, "Does Global Agriculture Need Another Green Revolution?" *Engineering* 4, no. 4 (Aug. 2018): 449–51, https://doi.org/10.1016/j.eng.2018.07.017.
32. Daisy A. John and Giridhara R. Babu, "Lessons from the Aftermaths of Green Revolution on Food System and Health," *Frontiers in Sustainable Food Systems* 5 (Feb. 2021), https://doi.org/10.3389/fsufs.2021.644559.
33. Paolo D'Odorico et al., "Food Inequality, Injustice, and Rights," *BioScience* 69, no. 3 (Mar. 2019): 180–90, https://doi.org/10.1093/biosci/biz002.
34. World Food Programme, "Climate Change and Nutrition: A Case for Acting Now," Sept. 2021, https://docs.wfp.org/api/documents/WFP-0000131581/download/?_ga=2.122677442.892078228.1734108882-1090219487.1734108882.
35. Zheng Liu et al., "Driving Factors on Accumulation of Cadmium, Lead, Copper, Zinc in Agricultural Soil and Products of the North China Plain," *Scientific Reports* 13, no. 1 (May 2023): 7429, https://doi.org/10.1038/s41598-023-34688-6.
36. Mariana Portela de-Assis et al., "Health Problems in Agricultural Workers Occupationally Exposed to Pesticides," *Revista Brasileira de Medicina Do Trabalho* 18, no. 3 (2020): 352–63, https://doi.org/10.47626/1679-4435-2020-532.

37. Milla F. Brandao Gois et al., "Impact of Occupational Pesticide Exposure on the Human Gut Microbiome," *Frontiers in Microbiology* 14 (Aug. 2023), https://doi.org/10.3389/fmicb.2023.1223120.
38. Md. Ariful Islam et al., "Chronic Effects of Organic Pesticides on the Aquatic Environment and Human Health: A Review," *Environmental Nanotechnology, Monitoring & Management* 18 (Dec. 2022): 100740, https://doi.org/10.1016/j.enmm.2022.100740; Vinay Mohan Pathak et al., "Current Status of Pesticide Effects on Environment, Human Health and Its Eco-Friendly Management as Bioremediation: A Comprehensive Review," *Frontiers in Microbiology* 13 (Aug. 2022): 962619, https://doi.org/10.3389/fmicb.2022.962619; Nian-Feng Wan et al., "Pesticides Have Negative Effects on Non-target Organisms," *Nature Communications* 16 (Feb. 2025): 1360, https://doi.org/10.1038/s41467-025-56732-x.
39. Statehouse News Bureau, "A Decade Ago, Toledo Lost Access to Its Water. Toxic Algal Blooms Are Still a Problem," Aug. 2, 2024, https://www.statenews.org/section/the-ohio-newsroom/2024-08-02/a-decade-ago-toledo-lost-access-to-its-water-toxic-algal-blooms-are-still-a-problem.
40. Tvisha Martin, Zoom interview with author, July 31, 2024.
41. Christine Sprunger, Zoom interview with author, July 31, 2024.
42. Gillian Stepek et al., "Human Gastrointestinal Nematode Infections: Are New Control Methods Required?" *International Journal of Experimental Pathology* 87, no. 5 (Oct. 2006): 325–41, https://doi.org/10.1111/j.1365-2613.2006.00495.x.
43. Qiaofang Lu et al., "A Review of Soil Nematodes as Biological Indicators for the Assessment of Soil Health," *Frontiers of Agricultural Science and Engineering* 7, no. 3 (Sept. 2020): 275–81, https://doi.org/10.15302/J-FASE-2020327.
44. Kamalika Saha, "Regenerative Agriculture: A Boost for Soil Health," *ASBMB Today*, Oct. 27, 2022, https://www.asbmb.org/asbmb-today/science/102722/regenerative-agriculture-a-boost-for-soil-health.
45. Megan H. Dixon et al., "Time of Arrival During Disease Progression and Humidity Additively Influence Salmonella Enterica Colonization of Lettuce," *Applied and Environmental Microbiology* 90, no. 9 (Feb. 2024), https://doi.org/10.1101/2024.02.16.580743.
46. US Food and Drug Administration, "People at Risk of Foodborne Illness," accessed Jan. 16, 2025, https://www.fda.gov/food/consumers/people-risk-foodborne-illness; Diane G. Newell et al., "Food-Borne Diseases—the Challenges of 20 Years Ago Still Persist While New Ones Continue to Emerge," *International Journal of Food Microbiology* 139 (May 2010): S3–15, https://doi.org/10.1016/j.ijfoodmicro.2010.01.021; Ramona A. Duchenne-Moutien and Hudaa Neetoo, "Climate Change and Emerging Food Safety Issues: A Review," *Journal of Food Protection* 84, no. 11 (Nov. 2021): 1884–97, https://doi.org/10.4315/JFP-21-141.
47. WHO, "Food Safety," Oct. 4, 2024, https://www.who.int/news-room/fact-sheets/detail/food-safety.
48. Josef Finsterer, "Triggers of Guillain–Barré Syndrome: Campylobacter Jejuni Predominates," *International Journal of Molecular Sciences* 23, no. 22 (Jan. 2022): 14222, https://doi.org/10.3390/ijms232214222.
49. Michael B. Batz, Evan Henke, and Barbara Kowalcyk, "Long-Term Consequences of Foodborne Infections," *Infectious Disease Clinics of North America* 27, no. 3 (Sept. 2013): 599–616, https://doi.org/10.1016/j.idc.2013.05.003.

50. US Government Accountability Office, "Food Safety: Status of Foodborne Illness in the U.S.," Feb. 3, 2025, https://www.gao.gov/assets/gao-25-107606.pdf; Christina Jewett, "Food Safety Jeopardized by Onslaught of Funding and Staff Cuts," *New York Times*, Mar. 19, 2025, https://www.nytimes.com/2025/03/19/health/food-safety-trump-fda-cutbacks-deadly-outbreaks.html; Christina Jewett, "Trump Administration Postpones Deadline for Companies to Record and Trace Tainted Food," *New York Times*, Mar. 21, 2025, https://www.nytimes.com/2025/03/20/health/food-safety-trump-administration-outbreaks-listeria-ecoli.html.
51. Alison Kafer, "Un/Safe Disclosures: Scenes of Disability and Trauma," *Journal of Literary & Cultural Disability Studies* 10, no. 1 (2016): 6.
52. CSG Justice Center, "The Accountability Gap: Unsolved Violent Crime in the United States," Tools for States to Address Crime, 2025, https://projects.csgjusticecenter.org/tools-for-states-to-address-crime/the-accountability-gap-unsolved-violent-crime-in-the-united-states/.
53. M. E. Falagas, K. Z. Vardakas, and P. I. Vergidis, "Under-Diagnosis of Common Chronic Diseases: Prevalence and Impact on Human Health," *International Journal of Clinical Practice* 61, no. 9 (Sept. 2007): 1569–79, https://doi.org/10.1111/j.1742-1241.2007.01423.x.

ALL WE HAVE TO LOSE (OR SAVE)

1. Dave Petley, Zoom interview with the author, July 3, 2024.
2. Jessica Goodkind et al., "Hmong Refugees in the United States: A Community-Based Advocacy and Learning Intervention," in *The Mental Health of Refugees: Ecological Approaches to Healing and Adaptation*, ed. Kenneth E. Miller and Lisa M. Rasco (New York: Routledge, 2004), 301–40, https://doi.org/10.4324/9781410610263-18.
3. MNHS Reference Staff, "Hmong-Americans in Minnesota: Overview," Gale Family Library, https://libguides.mnhs.org/hmong/ov, accessed May 29, 2025.
4. Pajau Vangay et al., "US Immigration Westernizes the Human Gut Microbiome," *Cell* 175, no. 4 (Nov. 2018): 962–72, https://doi.org/10.1016/j.cell.2018.10.029.
5. Danilo Ercolini et al., "From an Imbalance to a New Imbalance: Italian-Style Gluten-Free Diet Alters the Salivary Microbiota and Metabolome of African Celiac Children," *Scientific Reports* 5, no. 1 (Dec. 2015): 18571, https://doi.org/10.1038/srep18571.
6. United Nations, "Far from the Headlines: After 50 Years Refugees from Western-Sahara Are Still in Camps," UN Western Europe, Mar. 11, 2024, https://unric.org/en/far-from-the-headlines-after-50-years-refugees-from-western-sahara-are-still-in-camps/.
7. Ilse-Maria Rätsch and Carlo Catassi, "Coeliac Disease: A Potentially Treatable Health Problem of Saharawi Refugee Children," *Bulletin of the World Health Organization* 79, no. 6 (2001): 541–45.
8. Denis English, "Contemporary Western Medicine Has Its Pitfalls," *Annals of Neurosciences* 26, no. 1 (Jan. 2019): 1–2, https://doi.org/10.5214/ans.0972.7531.260101.
9. Kate Armstrong, "Consumer Vulnerability and the Transformative Potential of the Consumption of Complementary Alternative Medicine (CAM)," *Journal of Customer Behaviour* 16, no. 3 (Nov. 2017): 207–36, https://doi.org/10.1362/147539217X15071081721099.

10. Gashaw Hassen et al., "Clinical Implications of Herbal Supplements in Conventional Medical Practice: A US Perspective," *Cureus* 14, no. 7 (July 2022): https://doi.org/10.7759/cureus.26893.
11. U. Werneke et al., "Potential Health Risks of Complementary Alternative Medicines in Cancer Patients," *British Journal of Cancer* 90, no. 2 (Jan. 2004): 408–13, https://doi.org/10.1038/sj.bjc.6601560.
12. Richard L. Nahin and Barbara J. Stussman, "Costs of Complementary and Alternative Medicine (CAM) and Frequency of Visits to CAM Practitioners: United States, 2007," US Department of Health and Human Services, July 2009, https://www.cdc.gov/nchs/data/nhsr/nhsr018.pdf.
13. *Knowledge at Wharton* Staff, "Amway: Selling the Dream of Financial Freedom," *Knowledge at Wharton* (blog), May 5, 2011, https://knowledge.wharton.upenn.edu/article/amway-selling-the-dream-of-financial-freedom/; Lesley Fair, "It's No Longer Business as Usual at Herbalife: An Inside Look at the $200 Million FTC Settlement," Federal Trade Commission, July 15, 2016, https://www.ftc.gov/business-guidance/blog/2016/07/its-no-longer-business-usual-herbalife-inside-look-200-million-ftc-settlement.
14. Marcia Angell and Jerome P. Kassirer, "Alternative Medicine—the Risks of Untested and Unregulated Remedies," *New England Journal of Medicine* 339, no. 12 (Sept. 1998): 839–41, https://doi.org/10.1056/NEJM199809173391210.
15. Economic Research Service, "Food Security in the U.S.—Key Statistics & Graphics," USDA, updated Jan. 8, 2025, https://www.ers.usda.gov/topics/food-nutrition-assistance/food-security-in-the-us/key-statistics-graphics.
16. Chicago Food Equity Agenda, *Building a More Equitable Food System, Together*, https://www.chicago.gov/content/dam/city/sites/food-equity/pdfs/City_Food_Equity_Agenda.pdf, accessed May 29, 2025.
17. Christian A. Gregory and Alisha Coleman-Jensen, *Food Insecurity, Chronic Disease, and Health Among Working-Age Adults*, Economic Research Service, USDA, July 2017, https://www.ers.usda.gov/sites/default/files/_laserfiche/publications/84467/ERR-235_Summary.pdf.
18. Arline T. Geronimus, "Weathering the Pandemic: Dying Old at a Young Age from Pre-existing Racist Conditions," *Washington and Lee Journal of Civil Rights and Social Justice* 27, no. 2 (Spring 2021): 412–13, https://scholarlycommons.law.wlu.edu/cgi/viewcontent.cgi?article=1517&context=crsj.
19. Walter Leal Filho et al., "An Overview of the Interactions Between Food Production and Climate Change," *Science of the Total Environment* 838 (Sept. 2022): 156438, https://doi.org/10.1016/j.scitotenv.2022.156438.
20. Filho et al., "An Overview of the Interactions Between Food Production and Climate Change."
21. Matti Kummu et al., "Climate Change Risks Pushing One-Third of Global Food Production Outside the Safe Climatic Space," *One Earth* 4, no. 5 (May 2021): 720–29, https://doi.org/10.1016/j.oneear.2021.04.017.
22. D. H. Shugar et al., "A Massive Rock and Ice Avalanche Caused the 2021 Disaster at Chamoli, Indian Himalaya," *Science*, June 10, 2021, https://eprints.whiterose.ac.uk/id/eprint/175202/.
23. Jasper Knight, "Scientists' Warning of the Impacts of Climate Change on Mountains," *PeerJ* 10 (Oct. 2022): e14253, https://doi.org/10.7717/peerj.14253; Maximillian Van

Wyk de Vries et al., "Pre-collapse Motion of the February 2021 Chamoli Rock–Ice Avalanche, Indian Himalaya," *Natural Hazards and Earth System Sciences* 22, no. 10 (Oct. 2022): 3309–27, https://doi.org/10.5194/nhess-22-3309-2022.

24. Gregory Regelbrugge et al. v. Snohomish County (2018), slip opinion, Court of Appeals of Washington, Division 1, https://www.courts.wa.gov/opinions/pdf/777874.pdf.
25. Dan Shugar, phone interview with author, Oct. 17, 2024.
26. Y. Xu et al., "Forecasting Inundation of Catastrophic Landslides from Precursory Creep," *Geophysical Research Letters* 51, no. 15 (2024): e2024GL110210, https://doi.org/10.1029/2024GL110210.
27. King 5 Seattle, "Oso: Life After America's Deadliest Landslide," YouTube, Mar. 22, 2024, https://www.youtube.com/watch?v=LSNcedeotYk.
28. Noah Haglund, "Second Group Reaches $11.5 Million Oso Landslide Settlement," *Spokesman-Review*, July 22, 2018, https://www.spokesman.com/stories/2018/jul/22/second-group-reaches-115-million-oso-landslide-set/.
29. United Nations, "Greenwashing—the Deceptive Tactics Behind Environmental Claims," https://www.un.org/en/climatechange/science/climate-issues/greenwashing; World Economic Forum, "Climate Crisis May Cause 14.5 Million Deaths by 2050," Jan. 2024, https://www.weforum.org/press/2024/01/wef24-climate-crisis-health/, accessed May 30, 2025.
30. John R. Platt, "Christmas Island Bat, Last Seen in 2009, Confirmed Extinct," *The Revelator* (blog), Sept. 16, 2017, https://therevelator.org/extinction-christmas-island-bat/.
31. Dyani Sabin, "The Endangered Species List Is Full of Ghosts," *PopSci*, Feb. 26, 2019, https://www.popsci.com/extinct-species-still-on-endangered-list/.
32. Eli Clare, *Brilliant Imperfection: Grappling with Cure* (Durham, NC: Duke University Press, 2017), 6.
33. Clare, *Brilliant Imperfection*, 23.
34. Stephen Camarata, "Balancing Respect for Individuals, Human Rights, Neurodiversity, and Positive Behavioral Support in Intervention Research for a Spectrum of Autistic People," *Journal of Speech, Language, and Hearing Research: JSLHR* 65, no. 4 (Apr. 2022): 1607–9, https://doi.org/10.1044/2022_JSLHR-21-00660.
35. Naomi Shihab Nye, "Kindness," *Words Under the Words: Selected Poems* (Portland, OR: Far Corner Books, 1994).
36. Julia Watts Belser, "Disability, Climate Change, and Environmental Violence: The Politics of Invisibility and the Horizon of Hope," *Disability Studies Quarterly* 40, no. 4 (Dec. 2020), https://doi.org/10.18061/dsq.v40i4.6959.

BENT BODIES AND HIDDEN FIRE

1. Julie Cart, "Wildfires in January? Here's Why California Wildfire Season Is Worse," *CalMatters*, Jan. 8, 2025, http://calmatters.org/explainers/california-wildfire-season-worsening-explained/.
2. Susan Minnemeyer, phone interview with author, Nov. 20, 2015.
3. Matthew W. Jones et al., "Global Rise in Forest Fire Emissions Linked to Climate Change in the Extratropics," *Science* 386, no. 6719 (Oct. 2024): eadl5889, https://doi.org/10.1126/science.adl5889.
4. Emma Bowman, "A Decade on, the 'This Is Fine' Creator Wants to Put the Famous Dog to Rest," NPR, Jan. 16, 2023, https://www.npr.org/2023/01/16/1149232763/this-is-fine-meme-anniversary-gunshow-web-comic.

5. Piyush Jain et al., "Drivers and Impacts of the Record-Breaking 2023 Wildfire Season in Canada," *Nature Communications* 15, no. 1 (Aug. 2024): 6764, https://doi.org/10.1038/s41467-024-51154-7; NASA Earth Observatory, "Tracking Canada's Extreme 2023 Fire Season," Oct. 25, 2023, https://earthobservatory.nasa.gov/images/151985/tracking-canadas-extreme-2023-fire-season.
6. Brendan Byrne et al., "Carbon Emissions from the 2023 Canadian Wildfires," *Nature* 633, no. 8031 (Sept. 2024): 835–39, https://doi.org/10.1038/s41586-024-07878-z.
7. Natural Resources Canada, "Canada's Record-Breaking Wildfires in 2023: A Fiery Wake-up Call," Sept. 7, 2023, https://natural-resources.canada.ca/stories/simply-science/canada-s-record-breaking-wildfires-2023-fiery-wake-call.
8. Han Chen, Weihang Zhang, and Lifang Sheng, "Canadian Record-Breaking Wildfires in 2023 and Their Impact on US Air Quality," *Atmospheric Environment* 342 (Feb. 2025): 120941, https://doi.org/10.1016/j.atmosenv.2024.120941.
9. Brett Walton, "Wildfire Rampage in 2023 Injures Lungs in the Great Lakes," Michigan Public, May 14, 2024, https://www.michiganpublic.org/environment-climate-change/2024-05-14/wildfire-rampage-injures-lungs-in-the-great-lakes.
10. Zac Adelman, "Wildfire Smoke and Air Quality," Lake Michigan Air Directors Consortium, May 29, 2024, https://www.ladco.org/wildfire-smoke-and-air-quality/.
11. Statista, "Novel Coronavirus (COVID-19) Deaths by Country Worldwide 2023," https://www.statista.com/statistics/1093256/novel-coronavirus-2019ncov-deaths-worldwide-by-country/.
12. CDC, "End of the Federal COVID-19 Public Health Emergency (PHE) Declaration," Sept. 12, 2023, https://archive.cdc.gov/www_cdc_gov/coronavirus/2019-ncov/your-health/end-of-phe.html.
13. American Lung Association, "New Report: Chicago Metro Area Air Quality One of the Worst in Nation," news release, Apr. 23, 2024, https://www.lung.org/media/press-releases/chicago-worst-air-quality-state-of-air-report; Julian Relf, "Air Pollution Trends in Illinois," Institute of Government and Public Affairs, University of Illinois System, Aug. 27, 2024, https://igpa.uillinois.edu/wp-content/uploads/2024/08/Air-Pollution-Trends-in-Illinois-Policy-Spotlight.pdf.
14. American Lung Association, "New Report: Chicago Metro Area Air Quality One of the Worst in Nation."
15. Stephen J. Pyne, *The Pyrocene: How We Created an Age of Fire, and What Happens Next* (Oakland: University of California Press, 2021).
16. Stephen Pyne, phone interview with author, Oct. 9, 2024.
17. J. M. Blaut, "Colonialism and the Rise of Capitalism," *Science & Society* 53, no. 3 (1989): 260–96.
18. Jason W. Moore, "The Rise of Cheap Nature," in *Anthropocene or Capitalocene? Nature, History, and the Crisis of Capitalism*, ed. Jason W. Moore (Oakland, CA: PM Press, 2016), 1.
19. Sam Slater, "Charcoal in the Fossil Record," Research Communities, Springer Nature, Apr. 23, 2020, http://natureecoevocommunity.nature.com/users/254832-sam-slater/posts/66719-charcoal-in-the-fossil-record; Kristin Strommer, "Geologist Helps Confirm Date of Earliest Land Plants on Earth," *Oregon News*, Nov. 3, 2020, https://news.uoregon.edu/content/geologist-helps-confirm-date-earliest-land-plants-earth.
20. Juli G. Pausas and Jon E. Keeley, "Wildfires as an Ecosystem Service," *Frontiers in Ecology and the Environment* 17, no. 5 (2019): 289–95, https://doi.org/10.1002/fee.2044;

Anton Lokshin, Daniel Palchan, and Avner Gross, "Direct Foliar Phosphorus Uptake from Wildfire Ash," *Biogeosciences* 21, no. 9 (May 2024): 2355–65, https://doi.org/10.5194/bg-21-2355-2024.

21. J. A. J. Gowlett, "The Discovery of Fire by Humans: A Long and Convoluted Process," *Philosophical Transactions of the Royal Society B: Biological Sciences* 371, no. 1696 (June 2016): 20150164, https://doi.org/10.1098/rstb.2015.0164.
22. Steve Bradt, "Invention of Cooking Drove Evolution of the Human Species, New Book Argues," *Harvard Gazette*, June 1, 2009, https://news.harvard.edu/gazette/story/2009/06/invention-of-cooking-drove-evolution-of-the-human-species-new-book-argues/.
23. Francesco Berna et al., "Microstratigraphic Evidence of in Situ Fire in the Acheulean Strata of Wonderwerk Cave, Northern Cape Province, South Africa," *Proceedings of the National Academy of Sciences* 109, no. 20 (May 2012): E1215–20, https://doi.org/10.1073/pnas.1117620109.
24. Ran Barkai et al., "Fire for a Reason: Barbecue at Middle Pleistocene Qesem Cave, Israel," *Current Anthropology* 58, no. S16 (Aug. 2017): S314–28, https://doi.org/10.1086/691211.
25. Mercedes Conde-Valverde et al., "The Child Who Lived: Down Syndrome Among Neanderthals?" *Science Advances* 10, no. 26 (June 2024): eadn9310, https://doi.org/10.1126/sciadv.adn9310.
26. David W. Frayer et al., "Dwarfism in an Adolescent from the Italian Late Upper Paleolithic," *Nature* 330 (Nov. 1987): 60–62.
27. Frayer et al., "Dwarfism in an Adolescent from the Italian Late Upper Paleolithic," 61.
28. Lorna Tilley and Marc F. Oxenham, "Survival Against the Odds: Modeling the Social Implications of Care Provision to Seriously Disabled Individuals," *International Journal of Paleopathology* 1, no. 1 (Mar. 2011): 35–42, https://doi.org/10.1016/j.ijpp.2011.02.003.
29. Lorna Tilley, "The Bioarchaeology of Care," *Archaeological Record* 12, no. 3 (2012); Lorna Tilley and Tony Cameron, "Introducing the Index of Care: A Web-Based Application Supporting Archaeological Research into Health-Related Care," *International Journal of Paleopathology* 6 (Sept. 2014): 5–9, https://doi.org/10.1016/j.ijpp.2014.01.003.
30. Lorna Tilley, Zoom interview with author, Oct. 30, 2024.
31. Lorna Tilley, "Disability and Care in the Bioarchaeological Record: Meeting the Challenges of Being Human," in *The Routledge Handbook of Paleopathology*, ed. Anne L. Grauer (New York: Taylor & Francis, 2022), 462–63.
32. Mia Mingus, "You Are Not Entitled to Our Deaths: COVID, Abled Supremacy & Interdependence," *Leaving Evidence* (blog), Jan. 16, 2022, https://leavingevidence.wordpress.com/.
33. Reuters, "Americans Would Rather Be Dead Than Disabled: Poll," July 11, 2008, https://www.reuters.com/article/lifestyle/americans-would-rather-be-dead-than-disabled-poll-idUSN7B320259/.
34. James Greener, "Thomas Newcomen and His Great Work," *Journal of the Trevithick Society* 42 (2015): 64–127.
35. John Lienhard, "I Sell Here, Sir, What All the World Desires to Have—POWER," *Energy Laboratory Newsletter* (Feb. 1994): 3–9, https://engines.egr.uh.edu/sites/engines/files/talks/powerandfeedback.pdf.

36. Andreas Malm, "The Origins of Fossil Capital: From Water to Steam in the British Cotton Industry," *Historical Materialism* 21, no. 1 (2013): 15–68, https://doi.org/10.1163/1569206X-12341279.
37. Remi Kohler, "Nicolas Andry de Bois-Regard (Lyon 1658–Paris 1742): The Inventor of the Word 'Orthopaedics' and the Father of Parasitology," *Journal of Children's Orthopaedics* 4, no. 4 (Aug. 2010): 349–55, https://doi.org/10.1007/s11832-010-0255-9.
38. Leonard F. Peltier, *Orthopedics: A History and Iconography* (San Francisco: Norman Publishing, 1993), 21.
39. Berardo Di Matteo et al., "The 'Genesis' of Modern Orthopaedics: Portraits of Three Illustrious Pioneers," *International Orthopaedics* 37, no. 8 (Aug. 2013): 1613–18, https://doi.org/10.1007/s00264-013-1936-z.
40. Anne Borsay, "Disciplining Disabled Bodies: The Development of Orthopedic Medicine in Britain, c. 1800–1939," in *Social Histories of Disability and Deformity*, ed. David M. Turner and Kevin Stagg (New York: Routledge, 2007), 101.
41. Nora Groce, "Disability and the League of Nations: The Crippled Child's Bill of Rights and a Call for an International Bureau of Information, 1931," *Disability & Society* 29, no. 4 (Apr. 2014): 503–15, https://doi.org/10.1080/09687599.2013.831752.
42. Leanna Darlene Duncan, "Changing Bodies and Minds: 'Crippled Children' and Their Movement in the United States, 1890–1960," PhD diss., University of Illinois at Urbana-Champaign, 2021, https://hdl.handle.net/2142/110519.
43. Duncan, "Changing Bodies and Minds," 66.
44. Jeanne Hayes and Elizabeth "Lisa" M. Hannold, "The Road to Empowerment: A Historical Perspective on the Medicalization of Disability," *Journal of Health and Human Services Administration* 30, no. 3 (2007): 354.
45. Gelya Frank, "On Embodiment: A Case Study of Congenital Limb Deficiency in American Culture," in *Women with Disabilities: Essays in Psychology, Culture, and Politics*, ed. Michelle Fine et al. (Philadelphia: Temple University Press, 1988), 66–67.
46. Frank, "On Embodiment," 60.
47. Hayes and Hannold, "The Road to Empowerment," 361–62.
48. Audre Lorde, "The Transformation of Silence into Language and Action," *Sister Outsider: Essays and Speeches by Audrey Lorde* (1984; repr., Berkeley, CA: Crossing Press, 2007), 41.
49. Cristian Atala, Sebastián A. Reyes, and Marco A. Molina-Montenegro, "Assessing the Importance of Native Mycorrhizal Fungi to Improve Tree Establishment After Wildfires," *Journal of Fungi* 9, no. 4 (Apr. 2023): 421, https://doi.org/10.3390/jof9040421.
50. Paulo Freire, *Pedagogy of the Oppressed*, 30th anniversary ed., trans. Myra Bergman Ramos (New York: Continuum, 2005), 45.

COLONIAL CONTROL

1. Noel Preece, "Aboriginal Fires in Monsoonal Australia from Historical Accounts," *Journal of Biogeography* 29, no. 3 (Mar. 2002): 321–36, quote at 328, https://doi.org/10.1046/j.1365-2699.2002.00677.x.
2. Preece, "Aboriginal Fires in Monsoonal Australia from Historical Accounts," 328.
3. Livia C. Moura et al., "The Legacy of Colonial Fire Management Policies on Traditional Livelihoods and Ecological Sustainability in Savannas: Impacts, Consequences, New Directions," *Journal of Environmental Management* 232 (Feb. 2019): 600–606, https://doi.org/10.1016/j.jenvman.2018.11.057.

4. Stephen J. Pyne, *The Pyrocene: How We Created an Age of Fire, and What Happens Next* (Oakland: University of California Press, 2022), 70.
5. Stephen J. Pyne, "Fire in the Mind: Changing Understandings of Fire in Western Civilization," *Philosophical Transactions of the Royal Society B: Biological Sciences* 371, no. 1696 (June 2016): 20150166, https://doi.org/10.1098/rstb.2015.0166.
6. Moura et al., "The Legacy of Colonial Fire Management Policies on Traditional Livelihoods and Ecological Sustainability in Savannas."
7. Michela Mariani et al., "Disruption of Cultural Burning Promotes Shrub Encroachment and Unprecedented Wildfires," *Frontiers in Ecology and the Environment* 20, no. 5 (2022): 292–300, https://doi.org/10.1002/fee.2395.
8. Michael-Shawn Fletcher, Michela Mariani, and Simon Connor, "World-First Research Confirms Australia's Forests Became Catastrophic Fire Risk After British Invasion," *The Conversation*, Feb. 15, 2022, http://theconversation.com/world-first-research-confirms-australias-forests-became-catastrophic-fire-risk-after-british-invasion-176563.
9. Michael I. Bird et al., "Late Pleistocene Emergence of an Anthropogenic Fire Regime in Australia's Tropical Savannahs," *Nature Geoscience* 17, no. 3 (Mar. 2024): 233–40, https://doi.org/10.1038/s41561-024-01388-3.
10. Kirsten Vinyeta, "Under the Guise of Science: How the US Forest Service Deployed Settler Colonial and Racist Logics to Advance an Unsubstantiated Fire Suppression Agenda," *Environmental Sociology* 8, no. 2 (Apr. 2022): 134–48, https://doi.org/10.1080/23251042.2021.1987608.
11. US Department of the Interior, "The Department of Agriculture's Forest Service (FS)," June 19, 2015, https://www.doi.gov/protectnch/about-NCH/partners/fs.
12. Vinyeta, "Under the Guise of Science," 7.
13. Vinyeta, "Under the Guise of Science," 9.
14. Evan W. Kelley, "Forest Fire Control—a Form of Warfare," *Military Engineer* 25, no. 143 (1933): 407–10.
15. Robyn Schelenz, "How the Indigenous Practice of 'Good Fire' Can Help Our Forests Thrive," University of California, Apr. 6, 2022, https://www.universityofcalifornia.edu/news/how-indigenous-practice-good-fire-can-help-our-forests-thrive.
16. Elaine Scarry, *The Body in Pain: The Making and Unmaking of the World* (Oxford: Oxford University Press, 1987), 35.
17. Mary Johnson, "The 'Super-Crip' Stereotype: Sensitive News Topics," *FineLine: The Newsletter on Journalism Ethics* 1, no. 4 (July 1989), https://ethicscasestudies.mediaschool.indiana.edu/cases/sensitive-news-topics/the-super-crip-stereotype.html.
18. Maria Popova, "An Experiment in Love: Martin Luther King, Jr. on the Six Pillars of Nonviolent Resistance and the Ancient Greek Notion of 'Agape,'" *The Marginalian* (blog), July 1, 2015, https://www.themarginalian.org/2015/07/01/martin-luther-king-jr-an-experiment-in-love/.
19. Martin Luther King Jr., "The Christian Way of Life in Human Relations, Address Delivered at the General Assembly of the National Council of Churches," speech, Dec. 4, 1957, https://kinginstitute.stanford.edu/king-papers/documents/christian-way-life-human-relations-address-delivered-general-assembly-national.
20. Erin Arnold, phone interview with author, Nov. 1, 2024.
21. Grace C. Wright, Jeffrey Kaine, and Atul Deodhar, "Understanding Differences Between Men and Women with Axial Spondyloarthritis," *Seminars in Arthritis and*

Rheumatism 50, no. 4 (Aug. 2020): 687–94, https://doi.org/10.1016/j.semarthrit.2020.05.005.

22. Seneca, *Seneca: Moral and Political Essays*, ed. John M. Cooper and J. F. Procopé (Cambridge: Cambridge University Press, 1995), 34.
23. Rex Cleaver, "250 Years Since the 1774 Madhouses Act Gained Royal Assent," British Online Archives, https://britishonlinearchives.com/posts/category/notable-days/773/250-years-since-the-1774-madhouses-act-gained-royal-assent; Chantal Stebbings, "An Effective Model of Institutional Taxation: Lunatic Asylums in Nineteenth-Century England," *Journal of Legal History* 32, no. 1 (Apr. 2011): 31–59, https://doi.org/10.1080/01440365.2011.559119.
24. Adam Rutherford, *Control: The Dark History and Troubling Present of Eugenics* (New York: W. W. Norton, 2022), 54–55.
25. Facing History & Ourselves, "The Origins of Eugenics," Aug. 4, 2015, https://www.facinghistory.org/resource-library/origins-eugenics.
26. Nicholas W. Gillham, "Sir Francis Galton and the Birth of Eugenics," *Annual Review of Genetics* 35 (2001): 99.
27. Rutherford, *Control*, 69.
28. D. H. Lawrence, "I Would Build a Lethal Chamber," *The Art and Popular Culture Encyclopedia*, https://www.artandpopularculture.com/I_would_build_a_lethal_chamber_as_big_as_the_Crystal_Palace.
29. Cold Spring Harbor Laboratory, "Eugenics Record Office," https://www.cshl.edu/archives/institutional-collections/eugenics-record-office/, accessed May 30, 2025; Harry H. Laughlin, "Eugenical Sterilization in the United States (1922)," Embryo Project Encyclopedia, https://embryo.asu.edu/pages/eugenical-sterilization-united-states-1922-harry-h-laughlin.
30. "Buck v. Bell, Supreme Court of Appeals of Virginia, Brief for Appellee, September 1925," Document Bank of Virginia, https://edu.lva.virginia.gov/dbva/items/show/227.
31. Jasmine E. Harris, "Why *Buck v. Bell* Still Matters," Petrie-Flom Center for Health Law Policy, Biotechnology, and Bioethics at Harvard Law School, Oct. 14, 2020, https://petrieflom.law.harvard.edu/2020/10/14/why-buck-v-bell-still-matters/.
32. Rutherford, *Control*, 63; Harris, "Why *Buck v. Bell* Still Matters."
33. US Holocaust Memorial Museum, "How Many People Did the Nazis Murder?" https://encyclopedia.ushmm.org/content/en/article/documenting-numbers-of-victims-of-the-holocaust-and-nazi-persecution.
34. Rutherford, *Control*, 107.
35. "CDC Director Responds to Criticisms on COVID-19 Guidance," *Good Morning America*, Jan. 10, 2022, https://www.youtube.com/watch?v=gxZT7ra-oxs.
36. Marisa Kabas, "Disabled Americans Feel Abandoned by CDC. Now, CDC Is Desperate to Make Amends," *Rolling Stone*, Jan. 11, 2022, https://www.rollingstone.com/politics/politics-news/covid-cdc-disability-comorbidity-anger-1282759/.
37. Ben Matthews, "Camp Fire's Boundary Breaking-History," *Camp Fire* (blog), July 6, 2022, https://campfire.org/blog/article/camp-fires-boundary-breaking-history/.
38. A. E. Hamilton, "Putting Over Eugenics: Making It a Living Force Depends on Sound Application of Psychology and Sociology," *Journal of Heredity* 6, no. 6 (1915): 284.
39. Hamilton, "Putting Over Eugenics," 288, 287.
40. *Book of the Camp Fire Girls* (1958), 95.

41. UNDP Climate Promise, "The Climate Crisis Is a Health Crisis—Here's Why," June 6, 2022, https://climatepromise.undp.org/news-and-stories/climate-crisis-health-crisis-heres-why.

OUR DISABLED ECOLOGIES

1. Jean-Marc Cavaillon, "Once Upon a Time, Inflammation," *Journal of Venomous Animals and Toxins Including Tropical Diseases* 27 (Apr. 2021): e20200147, https://doi.org/10.1590/1678-9199-JVATITD-2020-0147.
2. Cavaillon, "Once Upon a Time, Inflammation."
3. Frederick W. Miller, "The Increasing Prevalence of Autoimmunity and Autoimmune Diseases: An Urgent Call to Action for Improved Understanding, Diagnosis, Treatment, and Prevention," *Current Opinion in Immunology* 80 (Feb. 2023): 102266, https://doi.org/10.1016/j.coi.2022.102266.
4. Frederick W. Miller, "Environment, Lifestyles, and Climate Change: The Many Nongenetic Contributors to the Long and Winding Road to Autoimmune Diseases," *Arthritis Care & Research* 77, no. 1 (2025): 3–11, https://doi.org/10.1002/acr.25423.
5. Yasmin Mohseni, Zoom interview with author, July 27, 2024.
6. Frederick Miller, Zoom interview with author, Nov. 1, 2024.
7. Angela Y. Davis, "Radical Perspectives on the Empowerment of Afro-American Women: Lessons for the 1980s (1987)," *Harvard Educational Review* 58, no. 3 (Aug. 1988): 352.
8. Sunaura Taylor, *Disabled Ecologies: Lessons from a Wounded Desert* (Oakland: University of California Press, 2024), 278.
9. James MacCarthy et al., "The Latest Data Confirms: Forest Fires Are Getting Worse," World Resources Institute, Aug. 13, 2024, https://www.wri.org/insights/global-trends-forest-fires.
10. Chantelle Burton et al., "Global Burned Area Increasingly Explained by Climate Change," *Nature Climate Change* 14, no. 11 (Nov. 2024): 1186–92, https://doi.org/10.1038/s41558-024-02140-w.
11. Chae Yeon Park et al., "Attributing Human Mortality from Fire PM2.5 to Climate Change," *Nature Climate Change* 14, no. 11 (Nov. 2024): 1193–1200, https://doi.org/10.1038/s41558-024-02149-1.
12. Matthew Nuttle, "Lives Lost in Lahaina: All of the People Who Perished in the Maui Fire," KITV Island News, July 31, 2024, https://www.kitv.com/news/lahaina/lives-lost-in-lahaina-all-of-the-people-who-perished-in-the-maui-fire/article_3b8e91ba-4122-11ee-a3f2-b73625af4d58.html.
13. Jennifer K. Balch et al., "The Fastest-Growing and Most Destructive Fires in the US (2001 to 2020)," *Science* 386, no. 6720 (Oct. 2024): 425–31, https://doi.org/10.1126/science.adk5737.
14. Nature Conservancy, "Bringing Indigenous Fire Back to Northern Australia," https://www.nature.org/en-us/about-us/where-we-work/asia-pacific/australia/stories-in-australia/bringing-indigenous-fire-back-to-northern-australia/, accessed May 30, 2025.
15. Jon Altman and Rohan Fisher, "The World's Best Fire Management System Is in Northern Australia, and It's Led by Indigenous Land Managers," *The Conversation*, Mar. 10, 2020, http://theconversation.com/the-worlds-best-fire-management-system-is-in-northern-australia-and-its-led-by-indigenous-land-managers-133071.

16. Michael-Shawn Fletcher et al., "Catastrophic Bushfires, Indigenous Fire Knowledge and Reframing Science in Southeast Australia," *Fire* 4, no. 3 (Sept. 2021): 61, https://doi.org/10.3390/fire4030061.
17. "Video: Northern Australia Fire Management with the Emissions Reduction Fund—DCCEEW," https://www.dcceew.gov.au/climate-change/emissions-reduction/accu-scheme/video-northern-australia-fire-management-with-the-emissions-reduction-fund, accessed May 30, 2025.
18. David Helvarg, "Using Fire to Fight Fire: California Tribes' Cultural Burns Restore Land and Keep Flames at Bay," *American Indian* (magazine of the Smithsonian's National Museum of the American Indian) 22, no. 2 (Summer 2021), https://www.americanindianmagazine.org/story/using-fire-to-fight-fire; National Park Service, "Indigenous Fire Practices Shape Our Land," last updated Mar. 18, 2024, https://www.nps.gov/subjects/fire/indigenous-fire-practices-shape-our-land.htm.
19. Lawrence Atencio, "Revitalizing Indigenous Stewardship with Cultural Burning," Amah Mutsun Land Trust, Summer 2020, https://www.amahmutsunlandtrust.org/nls20.
20. University of California Agriculture and Natural Resources, "How the Indigenous Practice of 'Good Fire' Can Help Our Forests Thrive," Apr. 12, 2022, https://ucanr.edu/blog/forest-research-and-outreach/article/how-indigenous-practice-good-fire-can-help-our-forests.
21. Kari Marie Norgaard, "Colonization, Fire Suppression, and Indigenous Resurgence in the Face of Climate Change," *YES!* (magazine), Oct. 22, 2019, https://www.yesmagazine.org/environment/2019/10/22/fire-climate-change-indigenous-colonization; USGS, "Embers of Wisdom: The Yurok Tribe and USGS Partnership in Culturally Prescribed Fire Management," Jan. 11, 2024, https://www.usgs.gov/news/featured-story/embers-wisdom-yurok-tribe-and-usgs-partnership-culturally-prescribed-fire.
22. Erica Tom, Melinda M. Adams, and Ron W. Goode, "Solastalgia to Soliphilia: Cultural Fire, Climate Change, and Indigenous Healing," *Ecopsychology* 15, no. 4 (Dec. 2023): 322–30, https://doi.org/10.1089/eco.2022.0085.
23. Melinda M. Adams, "Indigenous Ecologies: Cultivating Fire, Plants, and Climate Futurity," *Artemisia* 49, no. 2 (Oct. 2023): 20–29.
24. Tom, Adams, and Goode, "Solastalgia to Soliphilia," 322.
25. Glenn Albrecht, "Solastalgia: A New Concept in Health and Identity," *PAN: Philosophy Activism Nature* 3 (2005): 45.
26. Albrecht "Solastalgia," 55.
27. Donald I. Dickmann and David T. Cleland, *Fire Return Intervals and Fire Cycles for Historic Fire Regimes in the Great Lakes Region: A Synthesis of the Literature*, Great Lakes Ecological Assessment, 2002.
28. Marie Zhuikov, "The Stories Trees Tell," Sea Grant, University of Wisconsin, Dec. 12, 2022, https://www.seagrant.wisc.edu/news/the-stories-trees-tell/.
29. Illinois Prescribed Fire Council, "The Illinois Fire Needs Assessment," https://www.illinoisprescribedfirecouncil.org/assessment-summary.html, accessed May 30, 2025.
30. Erin G. Rowland-Schaefer et al., "Mapping Fire History and Quantifying Burned Area Through 35 Years of Prescribed Fire History at an Illinois Tallgrass Prairie Restoration Site Using GIS," *Ecological Solutions and Evidence* 3, no. 2 (2022): e12144, https://doi.org/10.1002/2688-8319.12144.

31. Tyler McMahon, "Effects of Controlled Burning and Grazing on Macromoth Diversity in an Illinois Tallgrass Prairie," master's thesis, Bradley University, 2024.
32. Glenn Albrecht, "Soliphilia," *Psychoterratica* (blog), Sept. 8, 2019, https://glennaalbrecht.wordpress.com/2019/09/08/soliphilia/.
33. Taylor, *Disabled Ecologies*, 252.
34. Robert Frost, "July 9, 1931, Letter to Louis Untermeyer from North Bennington, Vermont," *The Letters of Robert Frost to Louis Untermeyer* (New York: Holt, Rinehart and Winston, 1963), 209.
35. Shakeel Ahmad Bhat et al., "Impact of COVID-Related Lockdowns on Environmental and Climate Change Scenarios," *Environmental Research* 195 (Apr. 2021): 110839, https://doi.org/10.1016/j.envres.2021.110839.
36. Taylor, *Disabled Ecologies*, 105.